Animals of the African Year
the ecology of East Africa

Animals of the African Year
the ecology of East Africa
by Jane Burton

PHOTOGRAPHS BY JANE BURTON

BOOK CLUB ASSOCIATES
LONDON

Previous page:
Grant's gazelles graze lush pasture during the wet season, while rain almost blots out the distant hills.

A lion rests in the shade of a bush, guarding his nearby kill. Uncomfortable in the midday heat, he is plagued by flies and is panting heavily to keep cool.

The author gratefully acknowledges references to the many authors whose papers on various aspects of East African ecology have been published in the East African Wildlife Journal, volumes I–IX. Thanks are also due to the National Museum of Kenya, Nairobi and to the British Museum of Natural History.

Index compiled by Jacqueline Pinhey
Photographer's agent: Bruce Coleman Ltd.

This edition published 1974 by
BOOK CLUB ASSOCIATES
By arrangement with Eurobook Limited

Printed in the Netherlands by Smeets Offset, Weert

Introduction

East Africa is one of the most exciting places on earth. It contains the highest mountain in the continent of Africa, and the largest and deepest lakes. It is slashed across by the two long gashes of the spectacular Great Rift, pocked by hundreds of volcanoes, many small and long since dead, a few of immense size and beauty, and some still active today. Its old lava flows, its soda lakes, its sheer precipices and granite domes tell of the turmoil which stretched and tore and remodelled the earth's crust here, and produced some of the most spectacular and beautiful scenery in Africa.

The greater part of the area lies within five degrees north and south of the equator. There are no winters or summers, but two annual rainy seasons and two dry ones; and the day length scarcely changes throughout the year. On the equator the rains may be evenly spaced, separated only by mild droughts or by no drought at all. The countryside remains green, plants may flower or seed at any time, animals breed the year round. At the other extreme the drought may be severe and may last for the greater part of the year. Here the few storms of the annual rainy season produce fresh, spring-like greenery full of nesting birds, baby mammals, insects and flowers; but autumn comes almost before there has been any summer, and the wintry, apparent-deadness of leafless twigs and dry dust returns to the land.

The great variety of terrain of East Africa, combined with the varied climate, results in a large number of distinct habitats in which animals can live. These range from the parched, inhospitable deserts of northern Kenya through thorn scrub, rolling grassland, and many forms of tree savannah to the dripping forests and rain-soaked moorlands of the mountains. There are even patches of lowland rain forest in the extreme west of the area. In some of these habitats the profusion of animal life is more rich and varied than almost anywhere else on earth. Not only are there many species and vast numbers of large mammals, but there is a multiplicity of smaller life forms, each adapted to fill a place within its particular natural community.

Not long ago, most of East Africa was an untouched wilderness, teeming with this marvellous abundance of wildlife. After the remarkable killing-spree in which man indulged during the last century much of the wilderness and its wildlife no longer exists.

The dominating factor in the ecology of the area has been the activities of man, but it is not the purpose of this book to dwell on the changes he has brought about, though mention of human influence is unavoidable. Fortunately, enough of the wilderness survives to bear witness to the interlocking patterns of animal and plant relationships that have developed over many thousands of years. The complexity of these relationships is such that detailed coverage of the ecology of even one habitat might take a whole book, while adequate coverage of the whole of East Africa is a near impossibility. Instead of attempting to unravel the whole intricate web of life throughout the area, this book shows typical animals and plants interacting with one another, yet fitting perfectly into their chosen habitats, surviving the rigours of the yearly cycle of events in the climates in which they live. I hope thus to convey something of the over-riding unity and sense of purpose that is so evident in natural communities, unspoiled by civilized man.

Jane Burton
Albury
Surrey

The East African Region

flamingos
alopithecus
Deinotherium
Chalicotherium
giant baboon
Sivatherium
marabou stork

Africa, with its great range of climate and vegetation—from the Mediterranean fauna and flora in the north, through the perpetual green of the great forests, to the extreme conditions of life on the high mountains or in the deserts—has so many habitats and a fauna of such great interest and variety that it defies more than a superficial glance in a small space. It is convenient, therefore, to take a corner of the continent which is fairly representative of much of the rest of Africa, in the hope that a longer look at its ecology will be rewarded by a greater understanding of the continent as a whole. For East Africa has forests as well as grassland, and deserts as well as mountains, and the ranges of many of its most spectacular animals extend well beyond the political boundaries of the area.

East Africa is only a small part of the zoogeographical region known as the Ethiopian, which comprises the whole of Africa south of the Sahara. It includes the three large territories of Kenya, Uganda and Tanzania together with Rwanda and Burundi, and is bounded for the most part by natural barriers. To the east lies the Indian Ocean, to the west the lakes and mountains of the Western Rift. To the north is Ethiopia, a highland area with a somewhat different fauna, separated from Kenya by stony deserts. The Sudan and the Somali Republic, though separate, merge ecologically and geographically into northern Uganda and northern and eastern Kenya. To the south, Tanzania is clearly cut off from the southern Congo and from Malawi by Lakes Tanganyika and Malawi, but its boundary with Zambia is clearer on a map than on the ground. Between Tanzania and Mozambique the demarcation line is the Ruvuma River; but vast areas are geographically similar on either side of the river.

It is convenient to divide the continent of Africa roughly into quarters: North Africa, which is characterized by deserts; West and Central Africa, with their equatorial forests; South Africa which consists largely of high plateaux covered with dry grassland but also contains an area of desert; and East Africa, which, except for a narrow coastal plain, is mainly highland and is characterized by sub-desert steppes.

The distribution of the main vegetational zones neatly coincides with the distribution of rainfall throughout the continent. The area of highest rainfall is in West and Central Africa, and produces the great tropical forests of the Congo and Niger Basins. Around the forest edges the vegetation grades into woodland, into wooded savannahs and open grasslands in areas of seasonal rain, and finally into desert in the areas where rainfall is most sparse. The deserts also coincide with more temperate conditions in the northern and southern quarters of the continent, where hot summers alternate with warm winters. In the middle part of Africa it is hot all the year round, although in the highlands nights may be cool or even cold.

The history of exploration

A hundred years ago the continent of Africa held many secrets and challenged the most intrepid explorers. The Mountains of the Moon and the source of the Nile remained unknown to Europeans. Yet the discovery of Africa had begun about fifty centuries earlier when the Egyptians sent caravans southwards to bring back ebony, ivory and slaves to their ancient capital, Memphis. During the time of the Pharoahs the cultural influence of Egypt on the interior of Africa was very great and it can still be discerned today. Later the Phoenicians, travelling far beyond the Mediterranean in search of trade, sailed south along the Atlantic coast of Africa. About 600 BC a fleet took three years to sail around the continent, and in 450 BC Hanno took a vast fleet of sixty Phoenician galleys and 30,000 prospective settlers down the west coast almost to the equator. After the Phoenicians came the Greeks, followed by the Romans, who established their empire in North Africa and pushed southwards into Ethiopia and beyond. There was also some contact between China and Africa a thousand years ago. By that time Arab traders were well established on the east coast. They travelled to and fro on the monsoon winds, carrying in their dhows cloth, beads and ironware which they exchanged for ivory, slaves and raw metals. The Arabs rarely penetrated inland, relying on African traders to bring the goods and slaves to them, but large settlements grew up along the coast, and the Arabs

Previous page: Two million years ago East Africa was inhabited by a profusion of animals, many of giant size. Some – flamingos and marabous, zebras, antelopes and hyenas – had already evolved into much their present form. Some, such as rhinoceroses and baboons, were giants; their descendants, though much like them in form, are dwarfs in comparison. Others, then at the dawn of their history, were small in comparison with later forms: *Australopithecus*, with his ape-like jaws and small brain, was much shorter than modern Man. Other animals such as the chalicotheres, four-horned *Sivatherium*, the sabre-toothed tiger, and elephant-like *Deinotherium* were coming to the end of their span and have vanished entirely from the scene during the intervening millenia.

The clawed frog, a faunal link with South America and with the ancient supercontinent of Pangaea of which Africa once formed a major part.

were undisputed masters of East Africa until the arrival of the Portuguese in the fifteenth century.

It was mainly the lure of West African gold that first drew the Portuguese to explore Africa, though they also hoped to wrest the trade monopoly from the Arabs. During the first half of the fifteenth century they inched warily down the West African coast under the direction of Prince Henry the Navigator. Fifty years later they had established a safe harbour at Lagos, and with it the first of a chain of coastal forts intended to deter other European powers. Vasco da Gama's discovery of the Cape sea route to the Far East gave Portugal a monopoly in trade in West Africa, though the Arabs remained strong in the East. Portuguese strength was challenged in the seventeenth century, however, when other European nations, England, Holland, France, Sweden and Prussia, became interested in the profitable slave trade. By the end of the century the Portuguese had lost their brief supremacy and England had established hers.

Some twenty million Africans were deported as slaves and a further ten million died before embarking or on passage to the Americas, before opposition to slavery, started by the Quakers, gained momentum. As the era of slavery neared its end, people began to take a wider interest in Africa. The industrial revolution created a demand both for new sources of raw materials—such as rubber—and for new markets for its products. Scientific curiosity was aroused. James Bruce's explorations in Ethiopia encouraged the founding in 1788 in London of the "Association for promoting the Discovery of the Interior Parts of Africa". For although the Romans had colonized North Africa and the coasts had long been settled, the vast centre of Africa was unexplored. The obstacles were formidable: there were the natural barriers of deserts, swamps and mountains, as well as the hazards of disease and wild animals. The inhabitants were justifiably hostile, fearing strangers as harbingers of slavery.

In 1795 the first of the great explorers of Africa, Mungo Park, survived an exceedingly hazardous journey to find the River Niger. In 1822 Hugh Clapperton became the first white man to enter the great northern Nigerian market city of Kano. The Frenchman, René Caillie, was the first European to visit the legendary and ancient Timbuktu and return to tell the tale, and Heinrich Barth of Hamburg alone survived a three-man expedition to the region between Timbuktu and Cameroon. Thus France and Germany as well as Britain were drawn into the scientific exploration of Africa. At the same time the great wave of missionary exploration got under way, reaching its peak in the work of David Livingstone.

During the second half of the nineteenth century East and Central Africa were gradually explored by Europeans. Richard Burton and John Speke reached Lake Tanganyika in 1857; a few years later Speke, with James Grant, discovered the source of the Nile—the Ripon Falls, where the Nile leaves Lake Victoria. In the early 1880s Joseph Thomson who, like Grant, has a gazelle named after him, led a Royal Geographical Society expedition to East Africa. He climbed Mount Kenya—whose existence was still questioned—and travelled down the Rift Valley to Uganda.

By 1900 the heart of Africa was no longer a complete mystery. It had been explored and mapped, but still remained challenging to explorers, adventurers and missionaries. All over Africa the great European powers were scrambling to stake their claims to the continent in the final decades of the last century. The apportioning was a direct result of the discoveries of the previous half century, explorers creating spheres of influence for their home countries. Often the divisions established had little connection with the boundaries of indigenous tribes. Borders were marked on maps along latitudes and longitudes, or following mountain ranges, lakes or rivers—hence the neatness of East Africa on the map.

When the countries of Europe began to rule Africa, European–African relationships were transformed. Anthropologists intensified studies of African societies and customs, and the Africans, brought daily into contact with white settlers and administrators, began to adopt European customs to varying extents. Some tribes still resist the alien clothes, foods and customs; others have adopted European ways. High-speed jet airliners now deposit European tourists in parts of Africa which a hundred years ago could only have been reached after months of death-defying effort. Yet in another sense Africa is still the Dark Continent, and the end of African discovery is nowhere in sight.

In the twentieth century we know as much of the ancient cultures of Africa as the nineteenth century knew of its geography. No form of writing was invented in Africa, south of the Sahara, until very recently, so that all traditions are oral. In South Africa, spoken histories have been traced back over a thousand years, but archaeology offers the best opportunity to uncover the past, and archaeology in most of Africa is still in its infancy. Egypt has attracted the world's greatest archaeologists but whole areas in other parts of Africa have scarcely been touched.

From the writings of Arab explorers we know something of the grandeur and power of the ancient kingdom of Ghana and of the fabulous wealth of the kingdom of Mali which succeeded it ten centuries ago; but of the great civilization of Kush or Nubia we know almost nothing. The written Kushite language remains undeciphered, their temples, palaces and tombs undug. Descendants of the Kushites may have built the stone towns and terraces whose ruins can still be seen in East Africa, near Lake Natron and elsewhere, but the archaeological evidence here, as for most of the continent, is fragmentary and inconclusive.

The dawn of man in Africa

The remote ancestors of man in Africa have received more detailed study than his immediate ancestors. Africa may well have been the cradle of mankind, for pre-humans, the australopithecines, lived in southern Africa about two million years ago, at the dawn of the Pleistocene Period. Fossil remains of these ape-men have been found in South Africa, at Olduvai Gorge in the Serengeti Plains of north Tanzania, and at Baringo and Rudolf in Kenya. The large australopithecine *Zinjanthropus*, nicknamed Nut-cracker Man because of its small brain and very large and heavy jaws, appears to have lived in East Africa contemporaneously with another hominid, named by the finder, Dr L. S. B. Leakey, *Homo habilis*. It seems to have been a small, light plains-dweller, while Nut-cracker Man, larger and more powerful, was possibly a forest-dweller. Controversy exists over the validity of separating these ape-men and true men into the genera *Australopithecus* and *Homo*. It is apparent that both walked with an erect posture and that *Homo habilis* made and used crude tools. East Africa is at the centre of the area in which very early pebble tools have been found, suggesting that the earliest human culture may have arisen here.

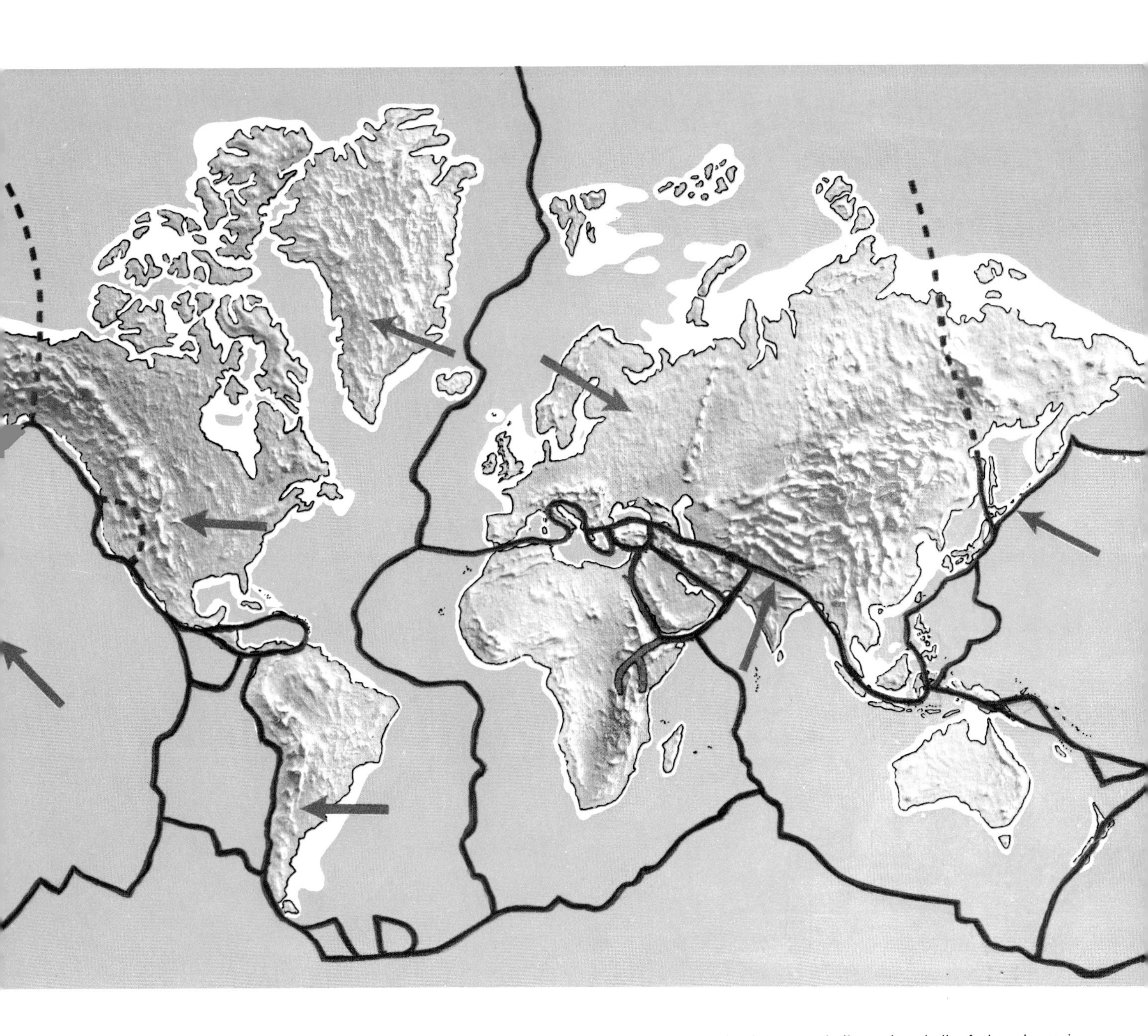

Many Earth scientists now believe the shell of the planet is broken into "plates", which are moving gradually in different directions as indicated by the red arrows. At the boundaries of plates, pressure has caused great mountain ranges to be thrown up – for example the Himalayas, the Andes and the Rocky Mountains. The Great Rift, shown here in red, is part of a series of submarine rifts along the shore of the Indian Ocean. Over five thousand kilometres long, it runs from south of the Zambezi Delta through East Africa and Ethiopia to the Red Sea and beyond. Two branches of the system run through East Africa, the Great Rift itself through Kenya and Tanzania, and the Western Rift to the west of Lake Victoria, through Rwanda and Burundi.

East African australopithecine and habiline man lived around the edges of lakes in temporary encampments. Their living floors are littered with bone fragments of fish, reptiles, birds and mammals as well as with naturally and artificially fractured quartz and obsidian stones. Outside the living area there are quantities of split and hammered bones of big mammals. The early man-like habilines probably lived partly by hunting small or slow creatures such as rodents, fish, snakes, frogs and tortoises, and partly by scavenging from predator kills. The australopithecines probably had a more vegetarian diet, like the modern gorilla, with whom they share certain skull characteristics. They probably ate bulbs, fruits, stems and nuts, as well as a small amount of meat, which they may have killed for themselves using bone or horn weapons.

Early man's animal neighbours

In the Pleistocene lake beds at Olduvai and at Olorgesailie near Lake Magadi, Kenya, an amazing variety of fossil animals contemporary with the hominids have been found. Many of the herbivores were enormous. For example, there was a giant sheep with a horn span of over a metre, and a pig as big as the modern rhinoceros. There was the elephant-like *Deinotherium*, with downward-curving tusks in the lower jaw and the tall straight-tusked elephant; there were the hefty, digging and browsing chalicotheres; and the elk-like *Sivatherium*, a relative of the giraffe, with a massive body over two metres long and two pairs of horns. Ostriches attained an even greater size than they do today, and there was a giant species of baboon and giant tortoises. There were various ancestral antelopes and gazelles, a big zebra and the little three-toed ancestral horse *Hipparion*, as well as vultures, flamingos, rhinoceroses and many other animals similar to those found in East Africa today. Preying on the herbivores were hyenas, the sabre-toothed tiger with its 18 to 20 centimetre fangs, and other carnivores of various sizes down to bat-eared foxes and genets.

The immense variety of African mammals then and now makes it surprising how little is known of their early history. Sedimentary deposits of the right age, in which remains might be found, are scarce in Africa. The earliest mammalian ancestors of today's fauna appear in lake beds in Egypt, dating from Eocene and Oligocene times, 50 to 30 million years ago, where the first diminutive elephants and the first monkeys and apes have been found. But south of the Sahara there are few fossil-bearing sediments earlier than the Miocene, and even these are rare except in East Africa. Here, however, on sites on the islands of Lake Victoria and near its shores in western Kenya and eastern Uganda, hundreds of fossil mammals that lived from 15 to 20 million years ago have been found, enabling palaeontologists to build up a picture of the fauna of the time.

Primates, normally rare among fossils, are exceptionally abundant here. There were three species of *Proconsul* apes, the largest as large as a gorilla and almost certainly related to it; an ancestral gibbon, and at least two galagos. Monkeys were rare at this time. Several kinds of insectivores have been found; these, too, are very rare from the Miocene elsewhere. Rodents were abundant, especially *Diamantomys*, an animal like a cane-rat, previously known only from the diamond deposits of South-West Africa. *Anthracotheres*, or ancestral hippopotamuses, and ancestral giraffes suggest that both these animals originated in Africa during the late Miocene. At that time there were no true antelopes—only chevrotain-like tragulids—but *Deinotherium* was already present, as were mastodonts. Carnivores were rare, but there were a great many reptiles—snakes, turtles, lizards and crocodiles.

Before the mammals

If mammalian ancestors are scarce, the reptile-like ancestors of the mammals are abundant in East African fossil-bearing rocks. The fauna of the East African Upper Permian is similar to the fauna of other parts of Africa at that time. Fossils of this age have been found in the Ruhuhu Valley near Lake Malawi in south-western Tanzania. In the same area fossils have been found from the Middle Triassic, about 210 million years ago—fossils which are of very great importance in furthering our knowledge of the prehistoric reptile fauna of the whole of Africa, because although abundant fossils have been found for earlier and later periods in South Africa, there is a gap in the sequence during the Middle Triassic, when conditions there were unfavourable for the laying down of fossiliferous strata. The Manda fauna of the Ruhuhu Valley neatly fills this gap. Here were reptiles of four main groups, including the crocodile-like ancestors of dinosaurs and of flying reptiles, crocodiles and birds, and the mammal-like reptiles from which true mammals evolved.

By the end of the Triassic the first dinosaurs were beginning to appear, and with them the first true mammals—small, shrew-like creatures that probably rummaged about among the leaf-litter for insects much as modern shrews do today. Gradually the dinosaurs increased in size until during the Upper Jurassic, 150 million years ago, the Earth was dominated by these gigantic animals. The biggest East African dinosaur was *Brachiosaurus*—indeed it was the largest land animal the world has ever known, weighing twenty times as much as a modern bull elephant, and able to browse up to 14 metres above the ground. Together with other giant herbivorous saurians, such as the plated stegosaurs, it must have demolished plants by the hundredweight, and was itself preyed upon by giant carnosaurs.

The fossil reptiles of East Africa have been satisfactorily documented, but so far there has been little to show of the fossil amphibians which gave rise to them. However, the sequence between the amphibians and the fishes is more completely documented in the fossil record than is the sequence between any other large groups. The ancestors of the four-legged land vertebrates were fleshy-finned fishes such as *Eusphenopteron*, of North America and Europe, which flourished in Upper Devonian times 350 million years ago. These fishes had primitive lungs and could crawl over land; they probably gave rise to the animals with well-developed limbs but fish-like tails such as *Ichthyostega*, the earliest known amphibian. The only surviving fleshy-fins today are the coelocanth of the Indian Ocean and the three genera of freshwater lungfishes. Lungfishes (page 96) are not in the direct line of descent between the rest of the fishes and the amphibians, but represent the kind of fish—capable of moving onto land and breathing by lungs—that led to the evolution of the amphibians.

The continent

Geophysicists have known for a long time that the Earth's continents consist of materials lighter than those composing the deeper layers of the Earth, and are floating upon them. The most favoured and acceptable theory about how the continents came to occupy their present positions is Wegener's theory of Continental Drift. Wegener pointed out that the continental shelves of the opposing continents could be fitted together very accurately. For example, the eastern bulge of South America could be made to fit neatly into the large bight of western Africa, as if the two had once been joined and had drifted apart. There is evidence, too, that Australia was earlier joined to south-east Africa. This theory of Continental Drift has now been confirmed along several lines of research.

As far as the animals are concerned, indications of continents drifting apart can be found in the distribution of several groups. One of the most spectacular is that of the lungfishes. The three lungfish genera are found today only in Australia, tropical Africa and tropical South America: if the theory is correct, in the heyday of the lungfish, these areas were more or less contiguous.

Other animals sharing this distribution are the large running birds, or Ratites: the emu and cassowaries of Australasia, the ostrich of Africa, and the rhea of South America. The African clawed frog has its South American counterpart in the pancake-shaped Surinam toad, while the side-necked terrapins, whose necks bend laterally in the horizontal plane, are also restricted to the southern continents: the snake-necked turtle in Australia, the helmeted water tortoise in Africa and the matamata in South America. Among the lower animals the odd-looking *Peripatus*, half worm, half insect, has a similar distribution today, and although it is no longer regarded as a direct link between worms and insects, as was once supposed, it nevertheless represents an archaic type in the lineage of insects.

The theory of Continental Drift is now generally accepted. Earth scientists are also confident that the continents are drifting because the surface of the Earth is broken up into a number of rigid plates of different sizes, all moving in different directions carrying the continents on their backs. What propels the plates themselves is still a matter for debate: convection currents welling up from the hotter underlying mantle of the earth may be moving them, or one plate becoming cold and slipping beneath another may cause currents that drag other plates to take its place. When plates collide, the great mountains of the world are thrown up. India colliding with Asia threw up the Himalayas, while the ocean that lay between them vanished. Similarly, the Urals were formed when Europe and Asia were in collision. But the mountains of East Africa were not created by the convergence of plates, for they do not occur at plate boundaries. They are paradoxical mountains that rose on huge domes, rather like gigantic bubbles welling up when heated from below, to form the great peaks and ranges that can be seen throughout the region today.

The Habitats

The Mountains

East Africa is one of the great volcanic areas of the world. Most of its higher mountain peaks are the cones or craters of extinct volcanoes. Some, such as Mount Elgon, and Mount Longonot in Kenya are still steaming. Oldonyo Lengai in southern Tanzania is gently active, occasionally spewing out soda lava so that its cone looks snow-capped. Astride the international boundaries of Rwanda, Uganda and Congo the Virunga volcanoes are still violently active.

Almost wherever we look we are reminded of the violent vulcanism of the past: by the volcanoes themselves, by basalt cliffs and still-naked lava flows, by boulders of pumice and obsidian, by soda lakes and the contorted strata of isolated hills; while steam jets and boiling springs show that volcanic activity seethes not far below the surface even today.

There are two main theories to explain why this extensive volcanic activity occurred in East Africa. Earth scientists are now confident that the surface of the Earth is broken up into a number of rigid plates, all moving in different directions, carrying the continents. At the boundaries between the plates major geological events take place and many of the world's volcanic areas are there. Other volcanic areas, of which East Africa is one, occur inland from plate boundaries. They may be either along old fault lines or weaknesses in the crust; or they may occur over *hot spots*, strong sources of volcanic energy stationary beneath the plates, in about twenty different parts of the globe. According to the second theory the African plate may have passed over one of these hot spots, which heaved up the north-eastern quarter of the continent into huge domes. Directly over the hot spot, volcanoes punched their way through to the surface, pouring out molten rocks from the Earth's mantle; and extensive rifts split the surface of the land. As the plate continued imperceptibly on its way, propelled by the same energy source, the first volcanoes were carried away from the hot spot and they cooled, while others sprang up violently to take their place.

The only violent volcanoes in East Africa today are two of the Virunga peaks, Nyamlagira and Nyiragongo. The most violent is Nyamlagira. The scene inside its mile-wide crater is one of utter desolation: black solidified lava encrusted with sulphur, sulphurous vapours steaming from clefts and smoke-holes, and two great cones of volcanic ash belching out smoke. Every few years the mountain roars and spews out a scarlet river of lava which rushes down the slopes at 12 kilometres an hour, sweeping aside all plant life until it plunges hissing into Lake Kivu, 15 kilometres away. During the eruption elephants leave the district in terror, but small animals are attracted to the fires at night, and predators in turn are attracted to the swarms of insects and rodents, dodging after them through the hail of rocks and ashes.

On cool, dewy mornings small steam jets can be seen gently puffing from vents in the Rift Valley wall. Later, as the air heats up, the steam vaporizes and the jets are invisible. In places, whole hillsides appear to be smoking in the early morning sunlight.

Previous page: The glistening white cone of Kibo, highest of the twin summits of Mount Kilimanjaro, Africa's highest mountain. As the snowfields and glaciers catch the sunset, a Masai giraffe, already in darkness, is silhouetted against the lower slopes.

The neighbouring and somewhat higher peak, Nyiragongo, also becomes active at the same time. This is one of the few volcanoes in the world that has a liquid lava lake inside its crater, and the lake boils and seethes when Nyamlagira erupts. The

A cold mountain stream cascades down the face of a basalt precipice in the Aberdare Mountains. Basalt is lava which cooled rapidly and, while solidifying, formed into characteristic hexagonal columns. Redwinged starlings, birds of the mountains, nest in coolness and safety on a ledge behind the fall. Giant groundsels sprout from the cliff.

other peaks are dormant now; Mikeno and the pointed Karisimbi, both over 4,000 metres high; flat-topped Visoke; Sabinio, Muhavura and Mgahinga; all were once explosive but are quiet now.

The three great isolated peaks of East Africa, Kenya, Kilimanjaro and Elgon, are huge extinct volcanoes. Mount Elgon may once have been the tallest of the three—it has the greatest circumference at the base—but today it is gently sloping and only 4,200 metres high. Its crater is enormous, the second largest in the world; only the spectacular Ngorongoro caldera in Tanzania is bigger. The floor of the Ngorongoro is grassland teeming with big game, but cold and altitude prevent more than a few specialized animals from browsing the alpine flora of Elgon's crater floor.

Mount Kilimanjaro, too, is still gently puffing steam. At almost 6,000 metres it is the highest mountain in Africa today. Elgon has no permanent snow, but Kibo, the highest peak of Kilimanjaro, is covered with snowfields and glaciers. Mount Kenya also bears glaciers and snowfields. This mountain is unique, for its single peak is a solid lava plug of an extinct volcano. The ancient crater rim has been almost eroded away, except for some fantastically-shaped remnants on the

Small montane forest trees thin out at about 3,000 metres. A mountain river, dashing downwards over old lava flows, passes a patch of falling bamboos that have flowered and died.

Right: East Africa includes three large countries, Kenya, Tanzania and Uganda, and two smaller ones, Rwanda and Burundi. The map shows these, with the mountains and lakes of the region and the areas set aside for wildlife as National Parks and Game Reserves.

The vegetational zones on East African mountains, from forest round the foot to snow on the highest summits.

outlying shoulders. It has been described as the most magnificent example of mountain architecture, a single exquisitely sculptured mountain peak standing isolated and aloof, 600 metres above its gently sloping pedestal of forest and moorland.

In the line of the Western Rift between Lakes Edward and Albert is another great mountain range, the Ruwenzori, the fabled Mountains of the Moon. Unlike most other East African mountains it is not volcanic, but is a fault block of very ancient greenstone heaved up from the basement rock which underlies the rest of the continent. The 3,300 metre Cherangani, an escarpment in the Great Rift, is another non-volcanic range. The Ruwenzori range is a high, domed plateau from which the peaks jut up in groups, the six central ones sparkling with perpetual snows and glaciers, the tallest, Mount Stanley, rising to 5,000 metres. Below the snow-line are numerous lesser peaks; the whole range is immensely rugged and nearly always shrouded in cloud. Ruwenzori has the highest rainfall of any East African mountain; great rivers gouge deep gorges as they hurtle down towards Lake Albert and the Nile. The exceptional humidity and the absence of a dry season gives year-round plant growth so that the upper slopes are clothed in most luxurious cloud-forest.

Kenya, Kilimanjaro and Ruwenzori are the only African mountains to bear permanent ice, for the Atlas in Morocco, although snow-covered, has no glaciers. Kilimanjaro still has the most extensive glaciers of the three, but they are steadily shrinking, as are glaciers elsewhere in the world, because rainfall no longer exceeds the rate of evaporation. The Ruwenzori glaciers are ice-caps, rather than ice-rivers; great rounded cornices covering the high cols and peaks, with huge icicles underpinning the exposed ridges. Today they do not extend below 4,200 metres, but at one time huge ice-caps covered all the East African mountains down to 1,500 metres, and great rivers of ice flowed down into the valleys. The retreat of the East African glaciers may be due to the world-wide tendency towards a warmer, drier climate.

The forests There are differences in rainfall on East African mountains but because of similarities of altitude, low temperatures and high radiation there are very similar belts of vegetation. The lower slopes, up to a height of 2,400 or 3,000 metres, are encircled by rain forests called *montane* forests. Montane forest is very similar to true tropical forest but has smaller trees, fewer lianas

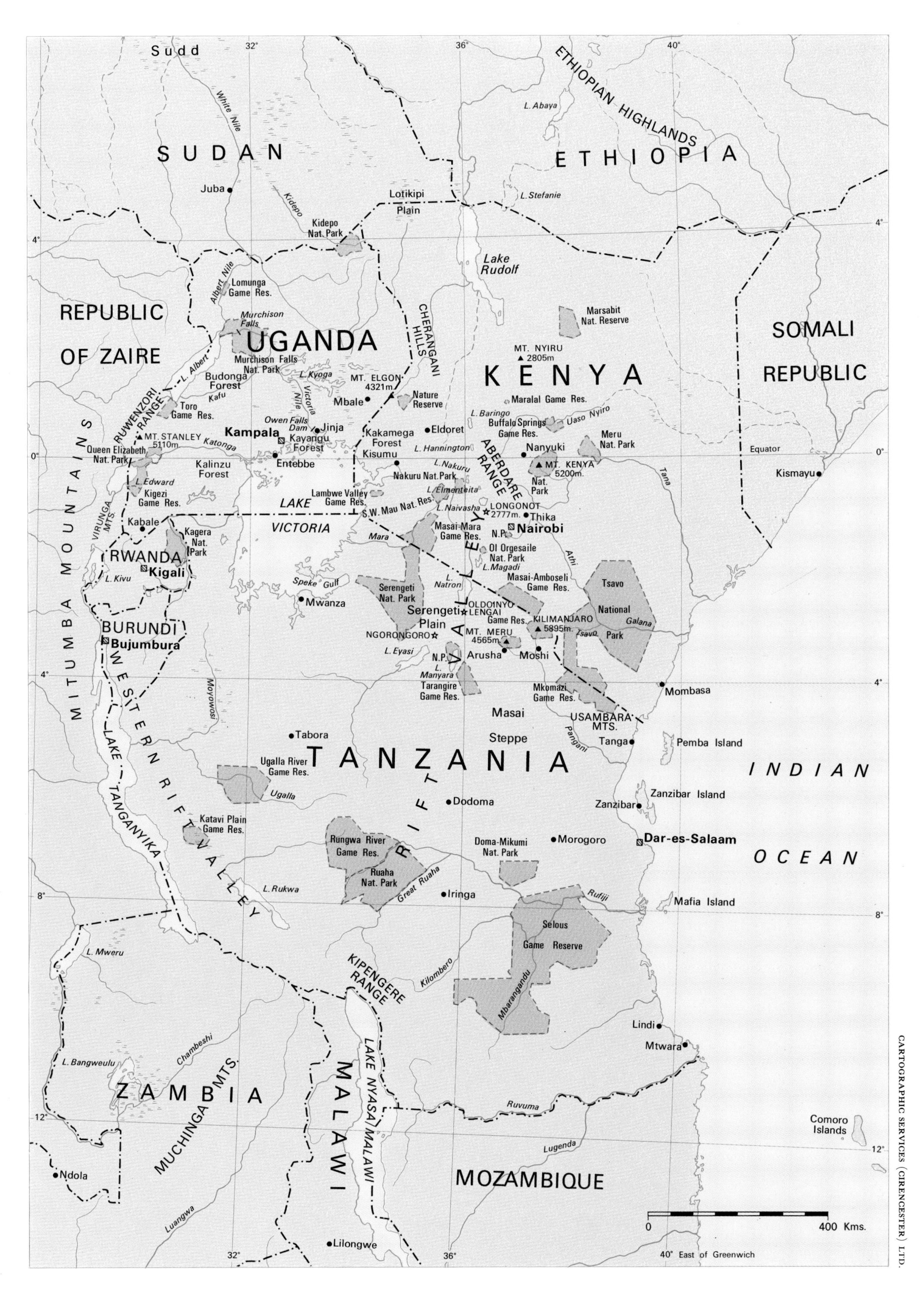

CARTOGRAPHIC SERVICES (CIRENCESTER) LTD.

Lichens thrive in the perpetual humidity produced by drizzle and mists at high altitudes. Streamers of beard lichen festooning the branches and twigs in montane forest are often more conspicuous than the trees' own leaves.

Many of the larger mammals of the forest are shy and elusive. The red duiker may be glimpsed in riverine and coastal forest and on mountains up to 2,400 metres. Above that height its place is taken by the very similar but slightly larger black-fronted duiker. The giant forest hog inhabits highland and rain forest areas and may sometimes be seen out at dusk at a forest edge waterhole during the dry season. The shyest and most elusive of antelopes is the bongo; it lives in the densest mountain forest and is very seldom seen. Both the great apes, gorilla and chimpanzee, may usually be glimpsed in their own forests, for they are not uncommon where they do occur; and they may even in time accept patient humans as their fellows.

The flame lily climbs among creepers and bushes in clearings and at the forest edge, its stems anchored by tendrilled leaf tips.

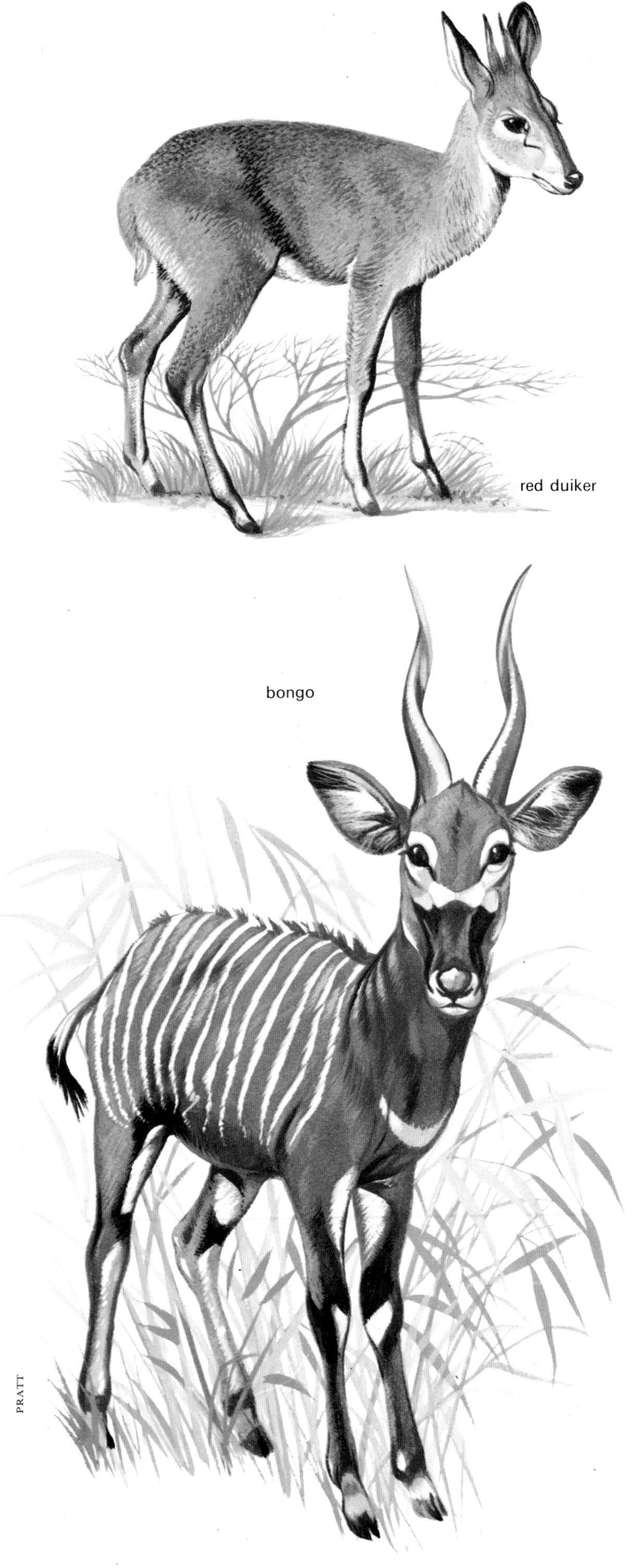

giant forest hog
gorilla
chimpanzee

and several gymnosperms—non-flowering evergreen trees. Stands of mature trees, of camphor, *Podocarpus*, cedar and red stinkwood, tower 30 to 40 metres above a wide variety of understorey trees. Beneath these great trees the ground may be clear but for a thick layer of leaf-mould and some small herbs. In other places, extensive thickets of wild bananas or bracken grow. The trees are festooned with parasitic creepers, beard lichens and tree orchids. Dwarf Kikuyu grass and pleasant flowering plants grow in the glades, together with groves of unpleasant giant stinging nettles.

At one time most of East Africa was covered by such forests, but today only about two per cent of the forest remains, at least in Kenya and Tanzania. Fragments of forest are scattered over most of Africa, and many of these remnant patches contain a remarkably similar fauna. High altitude butterflies are closely related to those of lowland forest, while many other kinds of animals found in East African forests are similar to species found in rain forests elsewhere. Some rain forest animals common elsewhere in Africa, such as some forest monkeys and hornbills, are now found only in the very west of Uganda. Here also may occur the tree pangolin, a very accomplished climber with a long prehensile tail; the brush-tailed porcupine, a smaller, spiny, climbing relative of the familiar, quill-covered crested porcupine; and the tiny dwarf galago which hops about in the tangled creepers and low bushes at the base of forest trees. Giant rats are more widespread in forest, but pottos and Fraser's scaly-tail or flying squirrel cannot now be found further east than Kakamega Forest, a tiny pocket of rain forest near Lake Victoria, and the last piece of true lowland forest left in Kenya. Even this remnant is now fast disappearing before the onslaught of the axes and chain saws of commercial forestry; soon the area will become swallowed up by the population pressure of the surrounding agricultural tribes.

Several East African peoples, notably the Kikuyu and Luo, traditionally live by shifting cultivation, clearing areas of the forest for their gardens and then moving on, allowing the luxuriant secondary growth to reinvade their plots and restore fertility. Today there is practically no more forest for such people to shift to when, after several years, the ground becomes unproductive. Many forests have been almost totally destroyed, the great trees turned into charcoal and the ground impoverished by the growing of maize, which gives back nothing to the soil. The people are innocent, but their population continues to increase. Not only the forest animals, but the Wanderobo, a truly forest people who live by hunting small animals and gathering wild honey, have little forest left to live in. Fortunately, many mountain forests are now included in national parks, for they are valuable catchment areas. But even forest reserves are not always sacrosanct from destructive resettlement schemes.

During the day montane forests, undisturbed by human settlement, are cool, dark places, full of the calls of birds and the crashing of monkeys in the upper canopy. An elephant snapping off a branch with a loud report may send a small bright-chestnut duiker scampering across a sunlit glade. There are many species of forest duikers, curious little hunch-backed antelopes with squat features and low head carriage, rather different from the more familiar bush duiker of open grassland. Forest duikers range from the tiny blue, only 30 centimetres high, to the relatively large and thickset yellow-backed. Even tinier and as secretive as the blue duiker is the suni, a graceful spike-horned forest antelope preyed on by the majestic crowned eagle. Largest of the forest antelopes is the bongo; in spite of its massive double-spiralled horns it moves easily among dense undergrowth. Duiker, suni and bongo are all mainly nocturnal, so they are very rarely seen.

Giant forest hogs live in small numbers on some East African mountains, in family sounders. They feed mainly at night, on grasses and herbs at the forest edge and on leaves and fallen fruit in the depths of the forest. The largest wild pig in Africa, their huge dark hulks may sometimes be mistaken for young buffaloes as they wallow in some forest glade.

In mountain forests there are no true seasons. Small antelopes and other small forest mammals may give birth, and birds nest, at any time of the year. Plants flower and fruit at any time. Insects, such as the huge metallic purple or green chafers that zoom onto forest flowers, are abundant. Rust-coloured shrew-rats find plenty of small invertebrates to eat as they patter, whiffling, over the leaf mould. Bird armies search the trees for small insects or follow the columns of safari ants that pass in an apparently unending stream across game paths and glades, under logs and round boulders. As many as 100,000 ants may pass in column, guarded on each flank by soldiers with formidable jaws. Anything that cannot get out of their path is carried along with them, whole or in

The blue-shouldered robin chat, a small thrush-like bird with a melodious song, lives in the undergrowth of lowland forests. It may join in bird parties which follow safari ants to snap up the insects they flush.

A column of safari ants on the march is an awesome sight—and a powerful ecological force. A hundred thousand tiny predatory insects stream in a close mass over the forest floor like some gigantic superpredator, overpowering and bearing along anything that cannot get out of the way.

shreds, from caterpillars to a recently-gorged python. Flycatchers, shrikes, bulbuls, barbets, ant thrushes, robin chats and small hornbills follow the ants and snap up any insects escaping from them.

Giant plants Above the montane forest is a narrow belt of bamboo, interspersed with a few of the same forest trees as grow lower down the slopes. Animals that inhabit the forests may also be found in the bamboo zone, although most of them do not eat the bamboo itself. Buffaloes, elephants and black rhinoceroses are all common, together with duikers, and the occasional bongo and forest hog. But there are no specialized bamboo-dwelling small mammals like the rats and bats that have evolved in the bamboo jungles of south-eastern Asia; and no animal that feeds almost exclusively on bamboo like the giant panda. As a habitat this belt is remarkably barren. The bamboos may be ten or more metres tall, with nothing but a few ferns and tiny herbs growing beneath them. About every thirty years they all flower simultaneously, producing quantities of seed, which may result in plagues of forest rats. After seeding the bamboos all die. When the seeds germinate the young plants form almost impenetrable thickets. Elephants browse the tops of the seedlings, but otherwise nothing is attracted to the bamboo zone, only to the forest trees within it.

On three of the Virunga volcanoes, however, and in the Kayonza Forest of Uganda, gorillas feed on bamboo at certain times of the year when young shoots are sprouting. Gorillas are entirely vegetarian, and eat a great variety of plants—lemonwood, wild celery, goosegrass, muskthistle, wild and cultivated bananas and even stinging nettles which they pick and eat, apparently without minding being stung. On the volcanoes Mgahinga, Sabinio and Muhavura, the mountain race lives mainly in the luxuriant *Hagenia* and bamboo forests around 3,000 metres. About 150 centimetres of rain a year produces thick ground cover and abundant forage and gorillas are most numerous and at home here. In the Kayonza, known as the Impenetrable Forest, rainfall is only about 100 centimetres but the constant low clouds and high humidity produce dense vegetation on the steep ridge sides, and swamps in the valleys. The gorillas, which may be of the lowland race here, move up onto the more open tops of the ridges by day, to feed and to bask in the sun if it breaks through the mist. At night they move down into the valleys to avoid the colder air of the heights.

Gorillas are not found outside dense humid forests but chimpanzees, also forest animals, can live in very dry, scrubby areas with few trees. They are not found on the Virunga volcanoes but the long-haired race is not uncommon in many forests in East Africa, especially in the Budongo Forest of Uganda. Chimpanzees are mainly vegetarians, but they sometimes eat insects and birds' eggs, and in Tanzania they stalk and kill young baboons, baby antelopes and birds. On the Ruwenzori range they live in montane forests up to 2,700 metres, where they climb trees with far greater agility than the heavier gorillas, but they also spend much time feeding and moving about on the ground.

Above the bamboos, from around 3,500 metres, occur heather forests and the giant groundsel zone. On ridge tops, or where the soil is poor or rocky, tree heathers abound. Their trunks and the ground are thickly cushioned with green, yellow and gold moss, and their branches are hung with the tattered grey beards of lichens. In the valley bottoms huge bogs are filled with moss and tall tussocks of sedge, with the occasional giant lobelia standing like an obelisk. On well-drained slopes, richer soils support small woodland trees, chiefly tree-sized St. John's

Giant groundsels near the summit of Mount Elgon, sending up their spikes of daisy-like flowers. These plants are unique to the East African mountains; among them grow clumps of everlasting daisies, a link with the flora of South Africa.

A natural rock garden of small red hot pokers, two species of dwarf everlasting daisies and other tiny alpines, flourishing above the giant groundsel zone among golden mosses and lichen-covered rocks.

worts with lovely yellow or orange flowers, and *Hagenias*. Beneath these grow brambles, wild celery, yellow alyssum and banks of everlasting flowers, pink and white and yellow and red. Here the first of the tree groundsels appear.

With increasing altitude the scene is gradually dominated by these giant groundsels. There are several species, but they all look much alike and all send up flowering spikes to about 5 metres. Sometimes they grow singly, sometimes in such profusion that they form groundsel forests. The plants have a thick trunk-like stem and great rosettes of cabbage-like leaves, each over half a metre long and silvered over with fine hairs. As the plant grows, the dead leaves remain attached to the trunk in a ruff, always dry in the centre, in which small birds roost at night. The flower spikes shoot up in a profusion of tiny daisy-like yellow flowers. Among the groundsels grow graceful torch lobelias, also with their leaves in rosettes, and with tall flowering spikes of pale blue. The massing of foliage in low dense rosettes, which is very common in alpine plants, reduces water loss. The covering of silvery hairs also holds in moisture and is a barrier against the strong ultra-violet radiation at these heights. Other species of tree St. John's wort also grow here, and there are acres of everlasting flowers over a metre high.

The extent and purity of the various vegetational zones varies from mountain to mountain. Among heather forests, those of the Ruwenzori are the most spectacular; in the dripping misty half-light, the lichen-festooned forests have a strange unearthly quality. Of bamboo forests, that on Mount Kenya is the most extensive, with purer stands. On Mounts Kenya, Kilimanjaro, and Elgon and on the Aberdares there are extensive areas of open moorland, rich in herbaceous plants such as saxifrages and a lovely scarlet-flowered wild gladiolus. Above about 4,000 metres on all the mountains, giant lobelias and groundsels start to thin out and everlastings are represented only by dwarf species covered with white woolly hair. Gradually the vegetation is reduced to mosses and lichens until just below the snow-line nothing can grow and no animals live permanently.

The alpine zone The climate of the heather and alpine zones is extremely harsh; it can be very hot indeed on cloudless days, while at night the temperature plummets down to give frosts before dawn. These are quickly dispelled in the first hours after sunrise, but after mid-morning clouds

The klipspringer is at home on mountain cliffs and rocky ledges up to the snow-line. At sunrise it stands on a rocky pinnacle, poised on its points, soaking up the sun's warmth to dissipate the coldness of the alpine night.

Mole rats are often very abundant in grassy meadows at high altitudes, tunnelling just below the surface and feeding on grass roots. They are one of the most important prey animals of the area, providing food for all predators from leopards to mole snakes. Their normal colour is chestnut brown, but black, white or piebald animals are by no means rare.

and mist begin to creep up the mountains and rain falls, sometimes as hail or snow. Nevertheless, many mammals are plentiful at these altitudes. Bushbuck, duikers, even eland, have been seen up to 4,000 metres. Buffaloes are abundant; they have plentiful grazing in forest glades and on the higher alpine slopes; trees to shelter among; mud for their wallows; and above all, peace, from lions and from man. At high altitudes they develop thick shaggy coats as a protection from the cold night air.

Among the high mountain crags live klipspringers, feeding on mosses and succulents. They have the compact build of the alpine chamois, and their elongated hooves, the texture of hard rubber, give them a sure-footed grip on the rocks. But the most typical animal of high mountains is the hyrax or dassie. It lives among outcrops and rocky boulders, running easily up and down smooth almost vertical surfaces, its rubbery pads kept moist by glandular secretion. If the hyraxes themselves are not visible their presence is apparent from shiny surfaces on rock faces which catch the sun like sheets of mica. These are formed by the deposits which drip down from the hyraxes' latrines. Hyraxes regularly use one or two spots for this purpose. During rain the lavatories are flushed out, so that fertilizer is carried down to the plants growing around the base of the rocks. The hyraxes feed on these plants, thus operating a small closed fertility cycle.

There are three genera and a dozen species of hyraxes, but in the field it is almost impossible to distinguish between them, they are all so alike in size, shape, colour and habits. Two genera live in rocky habitats, while the third lives in high forest trees. Rock hyraxes are extremely sociable, living in large colonies, like other ungulate herds. They are very vocal, too, calling and answering each other in high-pitched mewing and cackling calls which bounce and echo off the rocks. Both genera of rock hyraxes are entirely diurnal, but tree hyraxes are nocturnal and not as gregarious. By day, tree hyraxes sleep very high up in hollow trees or dense foliage, becoming active as darkness falls. They, too, are extremely vocal, yelling a series of very loud, harsh, creaking calls, terminating in drawn out piercing screams.

Both tree hyraxes and the two genera of rock hyraxes co-exist in scattered and isolated populations in East Africa. Generally it can be assumed that hyraxes among rocks are rock hyraxes and hyraxes in forest are tree hyraxes. But on the Ruwenzori and Mount Kenya the tree hyrax lives among cliffs and boulders in the upper treeless zone. Not only does it here occupy the niche which rock hyraxes fill elsewhere, but it has adopted the

The rock hyrax lives among jumbled boulders where caves and crannies provide a safe refuge from leopards and Verreaux's eagles, and insulation from the extreme climate.

habits of rock hyraxes, living in large colonies and coming out to feed and bask in the daytime.

Leopards are common at high altitudes. They feed on klipspringers, hyraxes, bushbucks and duikers. They have even been recorded visiting up to 4,500 metres on Mount Kenya to catch small rodents, possibly grass-eating vlei rats or striped mice. Mole rats are often very abundant in high grassy plateaux. These are very successful burrowing rodents with enormous orange incisors, and they occur on alpine moorlands up to at least 3,500 metres. They feed on grass roots, tunnelling through the soft ground and throwing up molehills. They are quite noisy underground, digging and chewing, and tapping either with the teeth or by stamping the hind feet. Several kinds of burrowing animals produce a drumming noise with their hind

Jackson's francolin is confined to the forest and bamboo zones of Mount Kenya and neighbouring highlands. Like other francolins it feeds by raking over the leaf litter and animal droppings to find small invertebrates. Each bird must turn, aerate and mix tons of surface soil and debris in its lifetime.

In the high mountains the augur buzzard preys chiefly on mole rats, watching for them from a lookout post or while soaring overhead. At lower altitudes it takes other kinds of rats, as well as the occasional lizard or snake.

feet, but the mole rat's signals are slow tap-tap-taps. Their underground progress is also marked by the heaving up of earth above surface runs and mole-hills, and by grass stems disappearing underground.

Mole rats fall prey at high altitudes to augur buzzards and tawny eagles by day, and to mountain eagle owls and servals at night, as well as to the occasional leopard. Mole snakes pursue them underground. So at no time of the day or night are they free from predation, above or below the ground; and yet they continue to be very numerous.

The most conspicuous small birds in alpine moorlands are the dark brown mountain chats that flit about among boulders and perch on clumps of everlastings, flirting their tails and scolding. They nest in holes under rocks or in earth banks where elephants have mined for salt. In places francolins and white-necked ravens are locally common. The rather rare but spectacular Verreaux's eagle, a coal-black, white-rumped bird of the mountains, feeds almost exclusively on hyraxes and perches high on lofty crags.

Alpine swifts nest on steep cliff faces almost up to the snow-line, but fly down 3,000 metres every day to the foot of the mountain to feed on the more abundant insects there. Various species of sunbirds occur at different levels, feeding on the nectar of lobelia and St. John's wort flowers, and weaving their tiny hanging nests from the wool of various alpine plants. At night they roost in the giant groundsel ruffs. The problem of heat loss during sleep must be an acute one for such tiny birds at high altitudes. Perhaps, like the humming birds of South America, they are able to avoid dying of exposure by becoming cold-blooded. At night, some humming birds fall into a torpid state during which their body temperature falls to about 5°C. They lose less heat in this state because the difference between body and air temperature is very much less than usual.

There are few other small animals at these heights. The high-casqued chameleon may be found among lichen-covered bushes, and surprisingly, frogs occur quite high up; the cold mountain streams may be full of tadpoles. Freshwater crabs are plentiful, too, which accounts for clawless otters visiting alpine bogs. There are no indigenous fishes—they have been unable to surmount the rapids—but introduced trout flourish in the mountain pools.

Different species of sunbirds occur at all altitudes from sea-level almost to the snow-line, wherever there are nectar-producing flowers. Some are confined to high altitude moorlands where they feed on giant groundsels, lobelias and red hot pokers; others live only in forest; some, such as this scarlet-chested one, are found in a variety of habitats. Their long, slender beaks are specially adapted for probing tubular flowers such as those of aloes and they are of prime importance in ensuring the cross pollination of the plants on which they feed.

This high-casqued chameleon is unusually well camouflaged among the silvery-pink flowers of everlasting daisies because it is in the process of sloughing. When not shedding its skin it is usually found among small shrubs or even on rocks where its bright green and reddish colouring blends with leaves, lichens and mosses.

The Great Rift

The Great Rift is a gash in the Earth's surface 5,600 kilometres long, running from south of the Zambezi Delta northwards through the highlands of East Africa and Ethiopia to the northern end of the Red Sea and beyond. It is part of a much greater series of submarine rifts along the floor of the Indian Ocean. Two branches of the system run through East Africa, the Great Rift and the Western Rift. At one time it was considered a very remarkable geographical feature and was presumed to be still widening, to become an ocean in another few million years. But it is now known to have widened by only about 9·6 kilometres in 20 million years, while the Red Sea widened 32 kilometres during the same period, and far faster widening has occurred in other places.

In its youth the East African rift was more than a kilometre deeper than it is today, but it has been filled in by the hundreds of volcanoes along its length. Nevertheless, some spectacular scenery remains. Vertical cliffs face each other, sometimes only a few hundred metres, sometimes sixty kilometres apart, across a flat plain dotted here and

Flamingos epitomize the Rift Valley soda lakes: nowhere else in the world are these magnificent birds concentrated in such numbers nor seen in such glorious settings. In the brief golden glow that precedes the sunrise, the lesser flamingos are feeding quietly, concentrating on filtering algae from the soupy water. But the greater flamingos, coming into breeding condition, are wading about, turning their heads smartly from side to side as they work themselves up into the state of intense communal excitement which culminates in the wings-out courtship display (page 150) and mating.

there with the rounded cones of extinct volcanoes. Sometimes it looks as if a strip of ground has simply dropped a few hundred metres into the earth. In places the cliffs run for many kilometres through otherwise flat bush country.

These cliffs are the home of rock hyraxes and klipspringers, leopards and Verreaux's eagles, like the cliffs and rocky places on the mountains. They are also the haunt of Schalow's wheatear, a very local Rift Valley bird; and of the mountain reedbuck which lives on the grassy slopes and scree at the foot of the cliffs. Baboons climb noisily up to the high cliff ledges for the night. Swifts, too, seem to roost high on these cliffs. They collect towards evening in vast wheeling, screaming mobs, tearing through the air above the cliff tops. Long charcoal-grey flight-feathers spiralling down testify to the loss of a few individuals as prey to peregrine falcons, but these rare raptors have no appreciable effect on the vast numbers of the swifts.

Vultures, particularly the griffons and the Egyptian vulture, roost and nest on cliff ledges. But the lammergeier is the true bird of great cliffs and gorges. In East Africa it is rare and very local, but a pair regularly roosts in the spectacular Hell's Gate Gorge near Lake Naivasha in Kenya. In the early morning they spiral up slowly and majestically, their long, narrow wings taking them skywards on the faintest up-draughts from the cliffs.

Soda lakes In prehistoric times, the floor of the Great Rift was filled with swamps and great lakes, but today all that remain of these are plains of white dust, deposits of diatoms, and a line of much smaller alkaline lakes. Some of these are fed by rivers, others by hot springs. The volcanic rocks of the escarpments and mountains on each side of the Rift Valley are exceptionally rich in carbonates. The water permeating through is charged with salts, chiefly sodium bicarbonate, which become concentrated in the lakes. Some have no outlets, but their evaporation rate is so high that, except in years of exceptional rainfall, they are actually shrinking. A few of the lakes are now almost dry: Lake Magadi is a flat, blinding-white waste of crystalline soda, 5 metres deep in places, overlying black mud. It is streaked here and there with pink algae, with pools of wine-red water, but from a distance it looks like an ice-rink, not a lake. It is the second largest expanse of pure washing soda in the world, and is extensively mined; but the inflow of soda from all its hot springs—over 400 tons a day—is greater than loss from commercial exploitation.

The hot springs of Magadi are the home of a remarkable dwarf cichlid, *Tilapia grahami*, which was forced to adapt to this extreme environment as the lake shrank and became more alkaline. The fish's ancestral forms grew larger and inhabited a much deeper, probably freshwater, lake; beds of their fossils have been found on escarpments above the present Lake Magadi. Today *Tilapia grahami* lives in water with a high specific gravity as well as high alkalinity and high temperature. Water temperatures in the springs vary from lukewarm to not far off boiling, but the fish are restricted to pools and lagoons with a temperature of about 38°C. Below this they become sluggish and cannot breed; only two degrees hotter and they die. They feed by browsing blue-green algae which grow on the gravel bottom, rasping the gravel clean in water of optimum temperature up to a distinct browse-line, where the water becomes too hot for them. But the algae grow most luxuriously in the hotter water and some fish are tempted by greener pastures to cross the browse-line, as the many corpses on the green bottom testify. Another similar tilapia species occurs in the hot springs of Lake Natron.

Lake Hannington is also fed by springs, which

Sheer vertical cliffs in the Rift Valley are the home of peregrine falcons and klipspringers. Baboons climb each night to high ledges; vultures and lammergeiers nest here. Among tumbled rocks at the foot live rock hyraxes, and on the grassy scree, mountain reedbucks.

Left: Schalow's wheatear flits among rocks in the Great Rift. Lichens are the first living things to colonize bare rocks, and can do so because they are dual plants made up of filamentous fungi and algae. The fungus part absorbs and holds moisture, and produces an acid which disintegrates the rock, allowing the plant to anchor itself. The algae manufacture food for the whole plant.

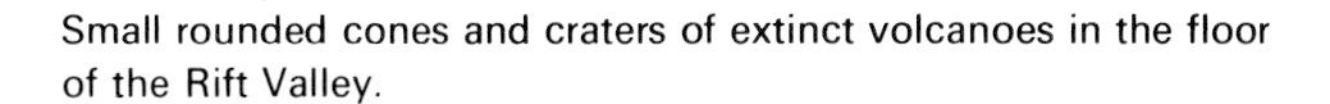

Small rounded cones and craters of extinct volcanoes in the floor of the Rift Valley.

The cooked bodies of dragonflies, moths, grasshoppers and other insects floating on the crystal-clear water of a boiling spring.

well up among rocks along the shore, crystal clear and only three or four degrees below boiling point. From these geysers the water streams down to the lake through channels in the rocks, cooling as it goes. Close to source the channels are encrusted with mineral deposits, but a little way down the shore they become overgrown with dense streamers of brilliant green algae, although the water is still very hot. In a nearby grassy meadow are deep pools, slightly less hot, with water deceptively still and clear, tempting animals to alight on the surface. White bones gleaming at the bottom show that birds have made the fatal mistake as have the dragonflies, water scorpions and other insects whose boiled bodies collect around the rims of the pools.

Only a few forms of animal life are able to live in the strongly alkaline waters of the Rift Valley

Lesser flamingos gathered to drink the fresh water by the flat, grassy meadow where Lake Hannington's boiling springs well and steam. The farther spring close to the lake shore maintains a channel through the flamingo band; the birds can drink but cannot paddle in really hot water.

lakes, but lakefly larvae, small water-beetles, water-boatmen, and copepod crustacea flourish in a soda solution. The murky waters are often thick with "water fleas", but are also a kind of living vegetable broth. Their high carbonate content, the warmth and sunlight, produce ideal conditions for the rapid multiplication of microscopic single-celled plants: diatoms, with beautifully sculptured siliceous skeletons, and blue-green algae, the most primitive plant occurring today. In some of the more alkaline lakes blue-green algae occur in such abundance that the water looks like pea soup—a fertile pasture for animals that can use it.

Flamingos Flamingos are the only large animals adapted for filtering suspended algae from the water. The lesser flamingo is the more highly specialized, and lives almost entirely on single-celled plants. The greater flamingo also filters algae, but in addition it takes in a lot of mud rich in organic matter, and sieves shrimps and small molluscs from the bottom. Since flamingos are the only birds making use of this food, the abundant algae are able to support them in vast numbers. Some three million flamingos live in the Great Rift, as many as occupy similar habitats throughout the rest of the world. They extract between them about 6 tons of algae per acre each year, yet the algae multiply at such a rate the water is never perceptibly less green.

The flamingos' habitat is as hot and shadeless as any desert, and flamingos need to drink as regularly as any other bird. They come once a day to the outlets from the boiling springs where the water is

relatively fresh and has cooled to about 75°C. They are just able to drink from the surface of this water although they cannot paddle in it. After drinking they go down to the edge of the lake to bathe, bobbing and fluttering in the water, then retire along the shore to preen. Below the springs there is constant excitement; all day birds are flying in to drink and splash while those that have bathed move on.

Large pythons among the rushes at the edge of a lake will occasionally grab a flamingo as prey, but the nature of their environment keeps flamingos otherwise fairly free from attack except from the air. Tawny and steppe eagles sometimes take them in flight with a powerful falcon-like stoop, and where the lake has no fish, fish eagles may feed almost exclusively on flamingos. Adult fish eagles that regularly feed on flamingos can be distinguished from fish-eating fish eagles by the colour of the feathers of throat and breast. Normally white, the feathers of flamingo-eaters are stained pink by the carotin in the flamingos' bodies.

But the chief predator of the flamingo is the marabou stork. It is primarily a scavenger, but around the soda lakes, where flamingos are not only exceedingly numerous but particularly vulnerable to the marabou's method of attack, it is a predator in its own right. A flamingo's one defence against the marabou is flight; once on the wing it is safe. But when it is paddling in deeper water take-off is not instantaneous. To become airborne a flamingo must first pull its legs up out of the water with a jump, then by half flying, half running, get enough lift to take it clear of the surface. When hunting, a marabou strides purposefully along the edge of the lake, between the flamingos and the shore, herding its prey into deeper water where take-off is most difficult. Suddenly it launches itself into the air and flaps out towards the flamingos which are now wading in a tight pack. Immediately they see this great dark shape heading towards them there is panic, and a brief confused struggle to take off. Those on the outside patter unimpeded across the open lake away from the marabou, but those on the inside of the throng must wait until the press has cleared a little before they can take off. Very often they all manage to get clear and the marabou returns to the shore to hound another section of the vast band. Finally, one bird on the inside fails to make a quick enough get-away, and the marabou pins it down in the water. Once a flamingo is wet, take-off is even more difficult, so the marabou has no need to make a swift clean kill.

Right: The marabou is a macabre-looking scavenger with an immense beak. It is also a predator and the chief enemy of the flamingo. Striding along the shore a marabou herds them into deeper water where take-off is most difficult. Suddenly it takes wing and the flamingos panic as they all try to get away at once. Inevitably one is unable to get clear in the confusion.

Far right: The tall, unlovely storks squabble mildly over the carcass while a spoonbill passes them, fishing as it goes, perhaps catching fish attracted to the underwater parts of the carcass.

Right, below: The flamingo's only defence from the marabou is to take off, pattering over the water; once airborne it is safe.

Adult steppe and tawny eagles are difficult to distinguish in the field. Both prey on lesser flamingos around the soda lakes.

The tiger beetle is an active predator, running swiftly about on sandy lake shores and pouncing on any other insect, such as this caddisfly, which it can grab in its large jaws. Perfectly camouflaged, it becomes virtually invisible as soon as it stops running.

The long-pincered earwig lives in burrows under logs or beneath boulders of jagged volcanic rock on the shores of soda lakes. Like the tiger beetle it is a voracious predator, but it emerges chiefly at night, whereas the beetle is active by day.

The shrill subterranean stridulations of mole crickets begin piercingly as dusk falls. The crickets tunnel through the grey lake-shore sand of volcanic grit and quartz grains, digging energetically with spade-shaped, mole-like forelegs.

Right: The avocet fishes by slicing through the water with upturned bill, catching various invertebrates and fish fry.

Other fauna of the soda lakes The fishless soda lakes support few other birds beside flamingos. Avocets and black-winged stilts find enough invertebrates by their own peculiar fishing techniques—avocets slicing through the water with upturned bill, stilts picking at the surface of water or mud. Dabchicks catch water-boatmen underwater; blacksmith and spurwing plovers, little stints and other migrant waders find food—aquatic insects—at the lake edge. Around the lake, over the sedges and yellow grass, mosquitos, midges and lakeflies may dance in clouds, following game like a moving halo, and these tiny flying insects are hawked by martins and swallows.

The grey sandy beaches of the soda lakes, strewn with jagged black lava rocks, look inhospitable deserts, yet some insects have adapted to this environment. Brown-and-yellow spotted tiger beetles run swiftly across the sand, pouncing on any other insect they can overpower. Another carnivore is the large, long-pincered earwig which hides under stones by day, even close to the water's edge, and comes out at night to feed. At dusk, the piercing stridulations of mole crickets accompany the twittering squeaks of the insectivorous bats that replace the swallows and martins at night.

The fish-eating birds of Lake Nakuru employ different fishing techniques.
Left: Grebes are the most perfectly aquatic of all flying birds. They pursue small fish under water, propelling themselves solely with their webbed feet. This African dabchick or little grebe is offering its mate a small *Tilapia grahami*.

Left, centre: Cormorants also pursue fish under water, but use their wings as well as their webbed feet for swimming. They have truly amphibious vision, but the waters of Lake Nakuru are so murky with suspended blue-green algae that birds must also be able to locate fish acoustically. After fishing, the white-necked cormorant holds its wings out to dry the sodden feathers.

Left, below: Spoonbills dash along, swishing their highly specialized bills from side to side, snapping up small fish which they locate by touch, or actively pursuing fish that they can see. Little egrets sometimes fish with them. The grey-headed gulls are partly scavengers, especially at marabou kills.

Right: The African wood ibis or yellow-billed stork wades along with its bill open and a wing out to cut surface reflections, sometimes feeling around with a foot to flush fish from hiding. The sense of touch in its bill is so acute and its reactions are so instantaneous that it can catch a fish that simply swims between its mandibles.

Far right: White pelicans fish co-operatively. From five to twenty birds swim along in a formation like a horseshoe with its open end foremost. Fish are driven along in front and concentrate in the horseshoe. At intervals all the pelicans plunge their bills into the water simultaneously, half-raising their wings at the same time.

Right, below: The smaller, pink-backed pelicans fish on their own, plunging the bill in whenever they see a fish near the surface. If the strike is successful, the pelican withdraws its bill carefully, allowing the water to drain out and trapping the fish in its pouch. The bird then throws back its head and gulps down the fish.

Lakes Elmenteita, Hannington and Nakuru have no endemic fish, but in the early 1960s *Tilapia grahami* was introduced into Nakuru from Lake Magadi as a mosquito-control measure. The fish multiplied so rapidly that their teeming descendants now attract thousands of fish-eating birds to the lake where none came before. There may be a thousand or more commuting pelicans there, a thousand resident pairs of cormorants and darters, uncountable numbers of dabchicks, herons of several species, fish eagles, scores of storks and spoonbills, hundreds of gulls and terns, and occasionally a few skimmers or an osprey. These birds all employ different fishing techniques and fish at various levels, but they are all feeding on the one species, apparently without over-fishing.

Feverthorns Near many of the soda lakes are stands of beautiful, tall feverthorn trees, in some places forming luxuriant forests. Their flat-topped crowns are not dense but are covered with a fine, lacy bright-green foliage through which the sun-

The sun filters through the canopy of a luxuriant forest of tall feverthorn trees, encouraging an undergrowth of creepers and shrubs, and picking out the conspicuous rump-pattern of a female defassa waterbuck.

light filters, so that an undergrowth of shrubs and creepers flourishes in the semi-shade. Colobus monkeys, turacos and trogons feed in the canopy and the trees are full of the calls of birds; wood-hoopoes, warblers, emerald-spotted wood doves, paradise flycatchers and bell shrikes. In the early morning and evening the horizontal sun cuts through the flimsy foliage and splashes the pale yellow trunks with patches of sulphur.

Feverthorns are found also along the banks of streams and rivers throughout the bush country, but they are most luxuriant around the lakes. The belts of woodland are rather narrow and many herbivores concentrate in and around them. Some animals inhabit main vegetation zones, pure grassland or dense forest, but very many prefer the varied vegetation found where two zones meet. At the edge of feverthorn woods, where dense undergrowth gives way to grassland, animals find a greater variety of food plants. They can bask in the sun, or keep cool in the shade. The woodland animals can creep among the dense undergrowth, while the plains animals can dash away in the open. Impala, waterbuck, warthog, Kirk's dikdik, black rhinoceros, elephants, bushbuck and vervet monkeys can all be seen most frequently at the edge of woodland, while buffaloes lie up in the forest during the day and come out into the grasslands to graze at night. The phenomenon of animals concentrating where two zones meet is known as *boundary effect.*

Kirk's dikdik is a dainty little antelope peculiar to dense bush country. It is often seen among thick shrubbery and undergrowth at the forest edge, where it is preyed upon by leopards and the large forest eagles.

Vervet monkeys often live in feverthorn forest, though not in the canopy. They may travel far from trees to forage on the ground but return to rest and groom each other in thick creepers and bush at the forest edge.

The black and white colobus monkey eats mostly leaves and lives exclusively in the canopy of tall trees in forested areas, montane as well as feverthorn forests.

The northern lakes The two northernmost lakes in the East African Rift, Lake Baringo and Lake Rudolf, are less alkaline than the southern soda lakes, and have a more varied fish fauna. Baringo contains quite large tilapia, mudsuckers and big catfish. Rudolf, a very much vaster stretch of water, also contains two species of very big and powerful predatory fishes. Nile perch, the largest non-marine fish in Africa, grows to about 2 metres in length and may live to over twenty years. Big Nile perch feed on nothing but other fishes such as big cichlids, smaller Nile perches, and tigerfish, a very large sharp-toothed characin which is almost as ferocious a predator as the Nile perch itself.

In most rivers and lakes of East Africa crocodiles have been almost wiped out for the sake of their skins. However, there are two places where they still enjoy something of their former abundance: the Victoria Nile in Uganda below the Murchison Falls, where they are protected to some extent in a national park; and Lake Rudolf. In this lake they are isolated by the surrounding deserts, but more importantly, the alkalinity of the lake water causes their skins to lose much of their commercial value, so they are not worth poaching.

In most localities, crocodiles feed mainly at night, though at Rudolf they are prevented from doing so by high winds that get up three hours after dark and blow until morning. Baby crocodiles feed mainly on insects—water scorpions and other aquatic species which prey on small fish, or land insects that drop onto the surface by mistake. From insects they graduate to small fish, and then to birds and mammals that fall into the water or can be taken when they come down to drink. Medium and large sized crocodiles also take Nile perch. They are therefore of great economic as well as ecological importance, particularly when they are very small or very big for they then feed largely on the predators of commercially important fishes.

African cichlid fishes are extremely prolific and fast growing. One species in particular, the Mozambique mouth brooder, is an important food fish and has been transported all over the tropical world. Other species have been introduced locally in East Africa into waters where they are not endemic, so the geography of cichlid species is often complicated. These cichlids owe their high rate of reproduction to a remarkable breeding behaviour. The females brood their eggs and young fry in their mouths, a method which ensures high success in brood rearing. In addition, the young fishes mature early. They start spawning at two to three months old, and are capable of multiplying a thousandfold in three months. Some species are quite small, but others may be 30 centimetres or more in length, and these are very valuable food fishes.

The eggs and fry of the Mozambique mouthbrooder are given a high degree of parental care, the female brooding them in her throat pouch. When still very young the baby fish are spat out into the water for short periods, but if danger threatens they rush back into their mother's mouth.

Golden cichlids show distinct sexual dimorphism, the male having the black belly. This species is one of 171 cichlid species endemic in Lake Malawi.

The Western Rift Lake Tanganyika, together with Lake Malawi (Nyasa) to the south, and Lakes Kivu, Edward and Albert to the north, lie in a very deep trench in the western arm of the African Rift. Lake Kivu is the most beautiful of this chain of lakes; it has deeply-indented fiord-like bays, formed when lava flows from the Virunga volcanoes dammed the valley which once drained northwards to the Nile. Its fish fauna, however, is limited and reminiscent of its previous nilotic connections, for though the lake now drains into Lake Tanganyika—which has a truly fantastic fish fauna—impassible rapids on the connecting Ruzizi River prevent new species from reaching Lake Kivu.

Both Lake Malawi and Lake Tanganyika are enormously deep; the latter has been sounded to a depth of 600 metres. Malawi is not quite so deep, but its bottom is still below sea level. In both, the lower layers of water are lifeless, for though they are rich in dissolved minerals these are not brought up to the surface by convection currents. However, the surface layers and lake margins support a remarkable aquatic fauna, including many endemic species such as a surface-skating, flightless caddis fly, aquatic snails, peculiar crustacea and the completely aquatic water cobra. They have also produced a great proliferation of fish species, especially among the cichlids, no doubt because of the enormous size of the lakes and their long geographical isolation. Lake Albert has only seven species of cichlid, of which two are endemic. By contrast, Lake Tanganyika has ninety species, all endemic except one, and Lake Malawi has 175 species all endemic except for four.

Bass and tilapia in Lake Naivasha support a profusion of fish-eating birds. This purple heron, only one of many heron species frequenting the lake, waited for three hours before the large tilapia it was watching swam near enough for a strike.

One of the most spectacular fishing birds in East Africa is the African fish eagle. It watches from a tall tree near the shore, then, flying low over the water, grabs a surfacing bass in its talons.

The tiny malachite kingfisher hunts from a low lake-side perch. It repeatedly whacks the tiny tilapia on the log before swallowing it.

Right: Islands of floating papyrus break off from the main mass and are blown across the surface of the lake. Like bamboos, papyrus is sterile as a habitat—nothing eats it and hardly anything can live among its dense tough stems.

Swamps and marshes Lake Naivasha is unusual among the Great Rift lakes in being freshwater. Fifty years ago it swarmed with a small indigenous fish, a top minnow. In 1926 a species of tilapia was introduced from the Athi River, and immediately began feeding on the top minnow. Later, bass from North America were also introduced, and between them these two very much larger species have apparently wiped out the top minnows. However, the big fish now support a profusion of fish-eating birds, including several, such as the purple and goliath herons, not generally found on the alkaline lakes.

Lake Victoria, the largest lake on the continent, and equidistant between the two arms of the African Rift, has a similar and wonderfully varied bird life. Like Lake Naivasha, it is dominated by a single plant species, a giant sedge, papyrus. The straight, leafless, triangular stems rise 5 metres high, with a mop of fine bracts at the top, and spring from a mat of dead stems and living roots which may be floating. The stems grow so thick and interwoven that a man has to cut a path through them. Scarcely any other plants can grow among papyrus, and few animals live in it. Only the extraordinary whale-headed stork or shoebill inhabits the vast papyrus swamps which surround the lake and stretch along the Nile basin to that greatest of all swamps, the Sudd, in southern Sudan. The shoebill hunts frogs and fishes deep in the swamps, and breeds there. When travelling from one swamp to another it soars very high in the thermals, so in spite of its great size it is very rarely noticed.

Underneath the floating islands of papyrus the water is dead, deprived of light and oxygen. In between it is often covered with great rafts of waterlilies, with lovely blue and purple flowers that open during the morning and close again by midday. Among the lilies there is far more animal life; in the clear pools between the leaves tilapia can be seen browsing or basking just beneath the surface; herons wade where the water is not too deep and kingfishers perch on the papyrus stems around the edges of the pools. Ducks and geese swim here to feed and wagtails run about on the lily leaves catching small flies. Dragonflies abound, and reed frogs set up their "pinkle-pinkle" chorus at dusk. Throughout the day land birds flock to suitable drinking points; even vervet monkeys visit the pools, climbing up the tall papyrus stems until these swing over with their weight to form convenient bridges across the channels. The

Where two vegetational zones meet, animal life is often particularly abundant. The purple heron does not live among the papyrus, but fishes at its edge. Other marsh birds, frogs, fish and aquatic insects congregate where papyrus meets waterlilies, where shade and cover are close to open spaces with plentiful food resources.

A purple gallinule holds a green waterlily pod in one foot while it feeds. A jacana and black crake circle hopefully, in case the bigger bird should abandon its pod half-eaten.

The extraordinary long toes and straight claws of the African jacana allow it to walk easily over floating vegetation, especially waterlily leaves. This bird has found a tiny frog, possibly a waterlily frog, which lives under the overlapping leaves, glueing them together with its spawn.

monkeys come in search of lily pods, which are a great delicacy. They walk out along a stout papyrus stem, lifting the lily leaves and peering into the water, then reaching in a long arm to pick the prized pod.

The most typical bird of the waterlily-papyrus marshes, and the best adapted to this habitat, is the African jacana, called the lily-trotter or lotus bird. Its very long toes and claws enable it to walk on the floating lily leaves without sinking in. Jacanas find all their food among the lilies, lifting the leaves with the bill and holding them folded over by treading on them with one foot. They find aquatic invertebrates, insect larvae, and tiny reed frogs; they also relish the lily pods, though they are not large enough to collect these easily. Even red-knobbed coots, which can dive down for the pods, have difficulty in breaking them from their anchoring stems. The only bird that can collect a pod with ease is the purple gallinule, a much heavier bird than the jacana, which has to run across the lily leaves to avoid sinking in at each step. When a purple gallinule brings a lily pod to the surface, it snips it neatly from the stalk, then, half-flying, it runs over the lily leaves to a firmer perch. There it holds the pod in the toes of one foot while it chops it to pieces. Smaller, would-be pod-eaters—jacanas, moorhens, Allen's gallinules and black crakes—circle around hoping to pick up some edible scraps. Sometimes they are lucky and the purple gallinule tires of the pod half-eaten. It seems deliberately to vary its diet, eating some pod then feeding on pink persicaria flowers or aquatic weeds before searching again for a pod. Thus all the pod-eaters get some occasionally, though none lives wholly on pods.

Papyrus swamps give way in places to reed and

Tiny reedfrogs abound among waterside vegetation. During the day they sit motionless for hours on a leaf or a reed, legs tucked in, exposed to the hot sun. At night they whistle in shrill chorus and feed on insects such as roosting damsel flies.

The sitatunga lives in marshes, usually feeding in the early morning or evening, hidden among the reeds. But this young bull was out feeding at midday on the lower stems of bullrushes.

A coypu, looking like part of the floating mat of vegetation, sits basking in the sun as red-knobbed coots feed.

bullrush marshes. Like the waterlily pools, these are full of birds: snipes, crakes, night herons, coucals, ibises, egrets, crowned cranes. Some weavers commonly build their spherical nests high up on tall reeds, and the thicket rat may later take over the old nests. Marsh mongooses hunt in family packs, crashing among the reeds as they pounce on frogs or birds. And in certain swamps sitatunga are common. Their greatly elongated hooves, which splay out under their weight, allow them to walk on sodden vegetation without sinking through. They move about the swamps in tunnels through the reeds, often up to the withers in water, walking on submerged vegetation. Their passage is marked by swaying bullrush heads or the loud squelching sucking noise of feet in and out of mud; but the animals themselves can remain completely hidden.

Surprisingly, in spite of such large areas of swamplands, which in former times were even more extensive, there are no native aquatic rodents in Africa. The shaggy rat lives in wet places, but not in true swamps, and so does the long-footed swamp rat. The cane rat, a relative of the porcupine, lives among elephant grass and other tall, rank vegetation in wet places, but although it can dive and swim if need be it is not truly aquatic. In some parts of East Africa, however, the coypu has become established. It is a large, South American, truly aquatic rodent with webbed hind feet, and was at one time farmed for its fur, known in the trade as *nutria.* When nutria went out of fashion many coypu were liberated in different parts of the world. In some East African lakes and swamps they flourished, feeding and giving birth to their precocious young on the floating mats of vegetation. Baby coypu may be taken by predators such as eagles, pythons and marsh mongooses, but an adult is a stout animal with very large orange incisors, and can in any case escape simply by slipping into the water. It will be interesting to see whether feral coypu will be successful in East Africa, and eventually take over a niche that,

By day hippos herd together in the water to keep cool and avoid sunburn. They have a profound influence on the ecology of rivers and lakes, since the amount of grass they consume on land by night is enormous, and much of this, converted into fertilizer, is released in the water during the day.

surprisingly, is not already occupied. Or it may be that the niche in fact does not exist, since no mammal has evolved to fill it.

Hippopotamuses A variety of amphibious animals is found throughout East Africa. Around Lake Victoria and other Uganda lakes the spotted-necked otter is common; in the rivers, streams and lakes of Kenya and Tanzania the larger clawless otter takes its place. In mountain streams of the Ruwenzori range and in the Kalinzu Forest in Uganda lives the solitary, nocturnal otter-shrew, hunting crustacea, otter-like, in small, fast-flowing streams. But in every suitable lake and river throughout the region, hippopotamuses are common.

Truly amphibious, hippos can swim at surprising speed, both at the surface and moving along the bottom, and they do so with extraordinary grace. They spend the day wallowing in the water, surfacing to breathe, with much snorting and ear-flicking. At night they come ashore to feed, half swimming, half wading through any floating mats of vegetation, breaking them up and maintaining channels which can be used by many smaller animals, such as ducks and geese, otters and marsh mongooses. They feed on the short grasses of water meadows, cropping the sward with the effect of lawnmowers. The males maintain territories ashore, with well-worn paths and landmarks beside which the owner of the territory sprays faeces and urine. At dawn the hippos trip down to the water again, daintily, in spite of their great bulk. During the day, they deposit more dung, in the water. On land the dung has social significance for the hippos, but since it is concentrated in a few places it is not very useful as a fertilizer. In the water it is of no importance to the hippos, but is a great environment enricher. Large herbivorous fishes may nibble some of the grass fibres, but more important, the nitrogen-rich dung fertilizes the green algae on which many kinds of fishes feed.

Riverine forest on the banks of the big permanent Mara river. Such forest is particularly rich in bird and mammal species, from monkeys and turacos to the very rare Pel's fishing owl, which used this fig tree as a daytime roost.

Hartlaub's turaco is the common turaco of highland and riverine forests. It feeds on fruit, running monkey-like along the branches. Though often unseen it draws attention by its raucous calls.

Riverine Forests The banks of rivers are very often bordered by strips of riverine forest, dense evergreen trees hung with creepers, through which elephants and hippopotamuses maintain broad paths on their way to and from the water. The very rare Pel's fishing owl roosts in a big old tree by day; at night it hunts fish and frogs in shallow reaches and backwaters. Wild fig trees grow in these forests. Their small latex-bearing green fruits sprout in bunches out of the trunks and branches, and ripen all at the same time to a delicious-looking reddish-orange. Within a few days the figs have been stripped by every fruit-eater within miles: monkeys and turacos, flocks of green fruit pigeons, hornbills, bulbuls, mousebirds and many others by day. At night fruit bats chew the figs, swallowing only the juices and spitting out pellets of dry fibres which patter on the forest floor like hail.

Bush pigs and red duikers are often common in riverine forest, and troops of monkeys may travel miles along forested water courses. Sun squirrels and bush squirrels, both arboreal, also live here. Like the narrow strips of feverthorn woodlands, this is a habitat particularly rich in species, for it contains not only forest-dwellers but also species that prefer the variety of vegetation found here.

Arboreal squirrels are far less common in East Africa than are the wholly terrestrial ground squirrels, but in riverine forest the bush squirrel sometimes occurs. It is preyed upon by forest eagles so any bird of prey passing overhead sets off its churring alarm call followed by a rapid dash to safety in a hollow tree.

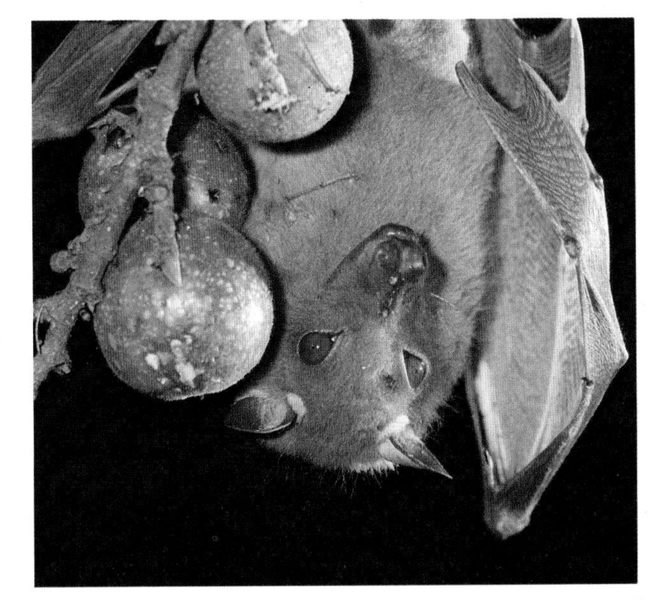

Wild fig trees in riverine forest may ripen during the dry season when other food is scarce, and provide a brief glut of fruit for monkeys and fruit-eating birds by day, and fruit bats by night.

The Bush Wilderness

The bushlands of East Africa are wildernesses of thorny thickets. They cover thousands of square kilometres of eastern Kenya and Tanzania, from the narrow coastal strip to the mountains. Here rainfall is often less than 50 centimetres a year, so the vegetation is adapted to near-desert conditions, and either taps water deep in the soil with very deeply-probing roots, or stores it in underground bulbs or tubers, swollen stems or succulent fleshy leaves. To conserve moisture, trees shed their leaves after the rains, and during the long dry seasons the bush looks dead. Grey and ochre are the predominant colours: bare twigs are grey; branches are festooned with grey-green leafless creeper stems; the dry earth is red; the scaly bark of trees is red or yellowish. In the dry season the bush is a hot and waterless place. Small river beds run only with dry sand; large permanent watercourses are few. In the rains, torrential downpours bring out a flush of ephemeral spring-like greenery and the rivers roar briefly. But the verdure and humidity soon give way again to bare twigs and drought. Yet this inhospitable region is the home of many big animals, well adjusted to the arid environment and at home in the wilderness.

The browsers The larger inhabitants of the bush are mainly browsers, for grass is often non-existent. Elephants, black rhinoceroses, giraffes, greater and lesser kudus, gerenuk, Grant's gazelles and dikdiks are the typical mammals of the bush. They are able to make the best use of this habitat because they are either nomadic, travelling long distances to water, or they can live entirely without drinking. They do not compete with each other, achieving ecological separation in their choice of terrain and by making use of the vegetation at all levels, from the ground up to about 6 metres. Giraffes are the tallest of all and the only animals that can feed on high foliage without having to push the tree over first. They may browse the tops of bushes so heavily that they make them stunted and flat-topped, while on the underside of taller trees they leave a browse line at 5 to 6 metres. In other places they keep small acacias and desert dates pruned into oval shapes. If the growth manages to get away above the giraffes' reach, however, the tree is clipped into the shape of an hour-glass.

Giraffes prefer flat country and browse chiefly on acacias. So do gerenuk. These exceptionally elegant antelopes, with their very long neck and limbs, can browse up to a height of 2·5 metres by standing upright on their hind legs with their forefeet among the branches. Kudus prefer steep hillsides and rough gullies and browse a wide variety of bushes and plants, including some generally regarded as poisonous. Tiny dikdiks browse the lowest levels of bush in all sorts of country, hilly or flat. And in all sorts of country elephants are equally at home. They need water regularly and in large amounts, so when there is nothing to drink and little but twigs to eat they seem out of place. But they are able to walk fast over long distances to get to water, and they find plenty of food by pushing over trees and stripping the branches. They are extremely catholic

The giant tuber of this elephant-foot vine protrudes from the soil, a reservoir of moisture and nutrients protected from desiccation by a corky bark. Even in the dry season its leaves stay green. The giant snails have succumbed either to drought or fire.

The greater kudu prefers boulder-strewn hillsides and rough gullies. A bull, with his twisted horns, throat fringe and majestic carriage is one of the most magnificent of antelopes.

The giraffe's long, prehensile tongue gathers and strips bunches of acacia leaves and soft young spines. Its pointed muzzle enables it to feed among the thorns without being pricked.

and adaptable in their choice of foods. So too are black rhinoceroses, which eat almost all the herbs and shrubs available in a habitat, even bitter latex-bearing euphorbias and plants, such as the thorn apple, that are highly toxic to other herbivores. They are very noisy feeders and on a quiet morning the sound of cracking branches and twigs being champed can be heard for several hundred metres.

Thorntrees Almost all the vegetation in the bush is thorny or prickly. Some acacia trees when young have particularly wicked eight-centimetre thorns; others combine long, straight, piercing thorns with shorter, curved thorns for holding. Some animals that feed on acacias are, therefore, adapted so that thorns are no problem. Many of the browsing antelopes have particularly small delicate muzzles that enable them to nibble leaves from among the thorns without getting spiked. Giraffes have long pointed muzzles and long, prehensile tongues for gathering bunches of leaves from between the thorns. Rhinoceroses eat the twigs, thorns and all, but do not look as if they enjoy them.

Some acacias have swollen thorns that look like the horns of Watutsi cattle, while others have gall-like swellings at the bases of the thorns. The best-known of the galled acacias is the whistling thorn, whose paired thorns sprout all along the twigs and branches from large round galls. The swellings are not true galls, which are growths produced by insects or mites. The whistling thorn normally bears these swellings, which are soft and at first have leaves. Later, as the galls harden and darken, the leaves disappear. Each hard old gall has a tiny round hole in it. The wind rushing over thousands of these tiny holes, amplified by the little sounding boxes of the hollow galls, produces an eerie moaning noise, which, though it is hardly a whistling, is the origin of the thorntree's name. Passing in and out through the tiny holes are small black ants which live in the galls, in a symbiotic relationship with the acacia. The ants gain some protection from the thorns, and in turn swarm onto and bite the muzzle of any animal browsing their tree. Impala are quickly moved on by the ants, but giraffe are as undeterred by ants as they are by thorns. Baboons also, are put off by neither, and will pick the galls and munch them up for the insect food they contain, spitting out the hard vegetable parts. The Abyssinian scimitar-bill also feeds on the ants, and is peculiarly well-adapted to do so. It is the smallest of the wood hoopoes, with a slender, finely curved, orange bill. It probes crannies in bark and among acacia flowers and will systematically work along the branches of a whistling thorn, extracting the ants, their eggs, larvae and pupae from the tiny holes in the galls.

Big old whistling thorntrees are also the home of acacia rats, the only truly arboreal rat of Africa. They build very large and conspicuous drey-like nests in the tops of trees, made of living twigs which they gnaw off for themselves. Berries, seeds and thorntree gum are their foods; they seem not to feed on the young galls or the ants. Acacia rats, and

The whistling thorn is the best known of the galled acacias. Left: The swellings, not true galls, are inhabited by tiny ants which try to protect the tree from browsers and in return themselves gain some protection from the thorns.

High among the branches of some big old whistling thorntrees are the drey-like stick nests of the acacia rat, the only truly arboreal rat in Africa.

The claw-like hind feet of the acacia rat, like those of other small arboreal animals, are reversible, enabling it to climb easily down branches head first.

Ground squirrels are very characteristic animals of the bush, and are often surprisingly bold, in spite of being prey to eagles, snakes, cats and other bush predators. When feeding they tripod on hind legs and tail to gain a wider view of their surroundings and guard against the danger of surprise attack.

The yellow-winged bat is one of the most conspicuous of insectivorous bats because it roosts in thorntrees and may fly before sunset. Very many other species roost in caves, hollow trees and roof spaces, and so are less often seen. Collectively they must consume tiny flying insects by the ton; they are themselves preyed on only by very specialized predators such as the bat hawk.

The Senegal bushbaby may live in thornbushes, but for obvious reasons prefers broad-leaved trees. It is exclusively nocturnal and bounds from branch to branch with prodigious leaps, feeding on insects, fruits and gum.

in other parts of the bush, tiny dormice, take the place of tree squirrels, which are not found away from forest trees. But ground squirrels, which almost never climb in trees, are very characteristic animals of the bush, running about in arid sandy areas with a curious jumping gait, their bushy tails undulating behind them. Patas monkeys, the true monkeys of the bush, have also taken to running on the ground. They are gaunt, brick-red animals, built like greyhounds and they can run very fast. They climb well, doing so to find fruits and seeds and to keep a lookout. But the usually small, almost always thorny and contorted trees of the bush are not suitable for monkeys to travel through at speed.

Larger, often isolated, thorntrees frequently have pairs of beautiful little yellow-winged bats hanging up to roost under the canopy. The bats are small and slate-grey, with enormous ears. Sometimes they fly before sundown; against the sun, their ears and wings glow orange. At night bushbabies spring from branch to branch, feeding on insects, flowers or fruits, and genets climb to hunt roosting birds, lizards, cicadas, crickets or whatever small prey they can find.

Birds of the bush Ground squirrels are often surprisingly bold and tame. They and dikdiks sometimes seem to be the only mammals about in the bush during the day. Patas are extremely shy and many of the herbivores either feed mainly at night or are very elusive. But if the mammals are mostly retiring, birds are often noisily conspicuous. Hornbills, barbets, bush shrikes, doves, go-away-birds, weavers, guineafowl, glossy starlings can all be located by their calls, so characteristic of the East African bush. Others, such as the many colourful bee-eaters, lilac-breasted rollers, dry-country kingfishers, fiscal shrikes and the shrike-sized pygmy falcon, perch conspicuously on bushes or dead trees, watching the ground for grasshoppers and other insects, small lizards or tiny snakes.

A small-spotted genet climbs in an elephant-damaged feverthorn tree. Genets are mainly nocturnal and carnivorous. Occasionally they will eat fruit, but generally they feed on any small animal—beetle, spider, lizard, rat, bird—which they can catch in the trees or on the ground.

Many bush-country birds perch conspicuously, on the lookout for insects or small reptiles. The aggressive fiscal shrike usually catches insects on the ground and returns to its lookout post—here the spire of a tree euphorbia—to demolish its prey.

The carmine bee-eater hawks a variety of passing insects—butterflies, small grasshoppers and beetles as well as bees—on short flights from its perch.

Right: The lilac-breasted roller can be wonderfully acrobatic on the wing, chasing a flying locust or mobbing a passing bird of prey. It is large enough to take the biggest insects, such as dung beetles, and swift enough to catch skinks and even small birds.

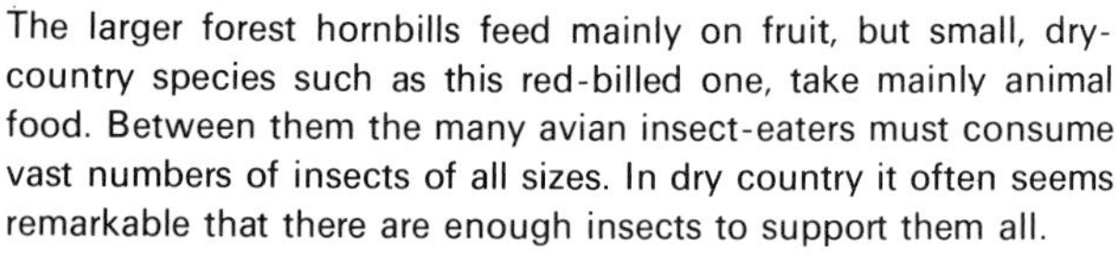

The larger forest hornbills feed mainly on fruit, but small, dry-country species such as this red-billed one, take mainly animal food. Between them the many avian insect-eaters must consume vast numbers of insects of all sizes. In dry country it often seems remarkable that there are enough insects to support them all.

Insects Many of the insects of the bush are beautifully camouflaged. The same species of small grasshopper may be red on red earth, grey on tufa-covered rocks near hot springs, mottled on stony ground. A very quaint, long-headed, green or buff-coloured grasshopper is found where grass is permanent. Among thorntrees, small cicadas are pale green to match the leaves, large ones mottled grey to match the lichen-covered bark. Various insects mimic acacia thorns: a moth at rest furls its forewings to resemble a straight thorn, while the hind wings are folded down to look like the swelling at the thorn's base. The very large pale green caterpillars of a saturnid moth are ornamented all over with beautiful small silver spikes that mimic thorns and also resemble the acacia's leaves.

In an attempt to escape predation from the insectivorous birds, some insects are superbly camouflaged. In a world dominated by spiky bushes, many insects achieve concealment by mimicking thorns. This caterpillar of a large saturnid moth, *Gynasa maja*, feeds on acacia.

Termites Termites are almost everywhere and are as numerous as leaves in the bush. They occur whereever there are trees, for they feed on dead wood. In the bush they consume all the acacias which have been knocked over and killed by elephants. They can eat wood because they have in their intestines microscopic protozoa which break down cellulose, turning it into sugar. Without these protozoa, termites would starve to death because they cannot digest wood on their own.

Many kinds of termites build large mounds, tall red towers which may reach a height of over 4 metres, and are a very striking feature of the bush.

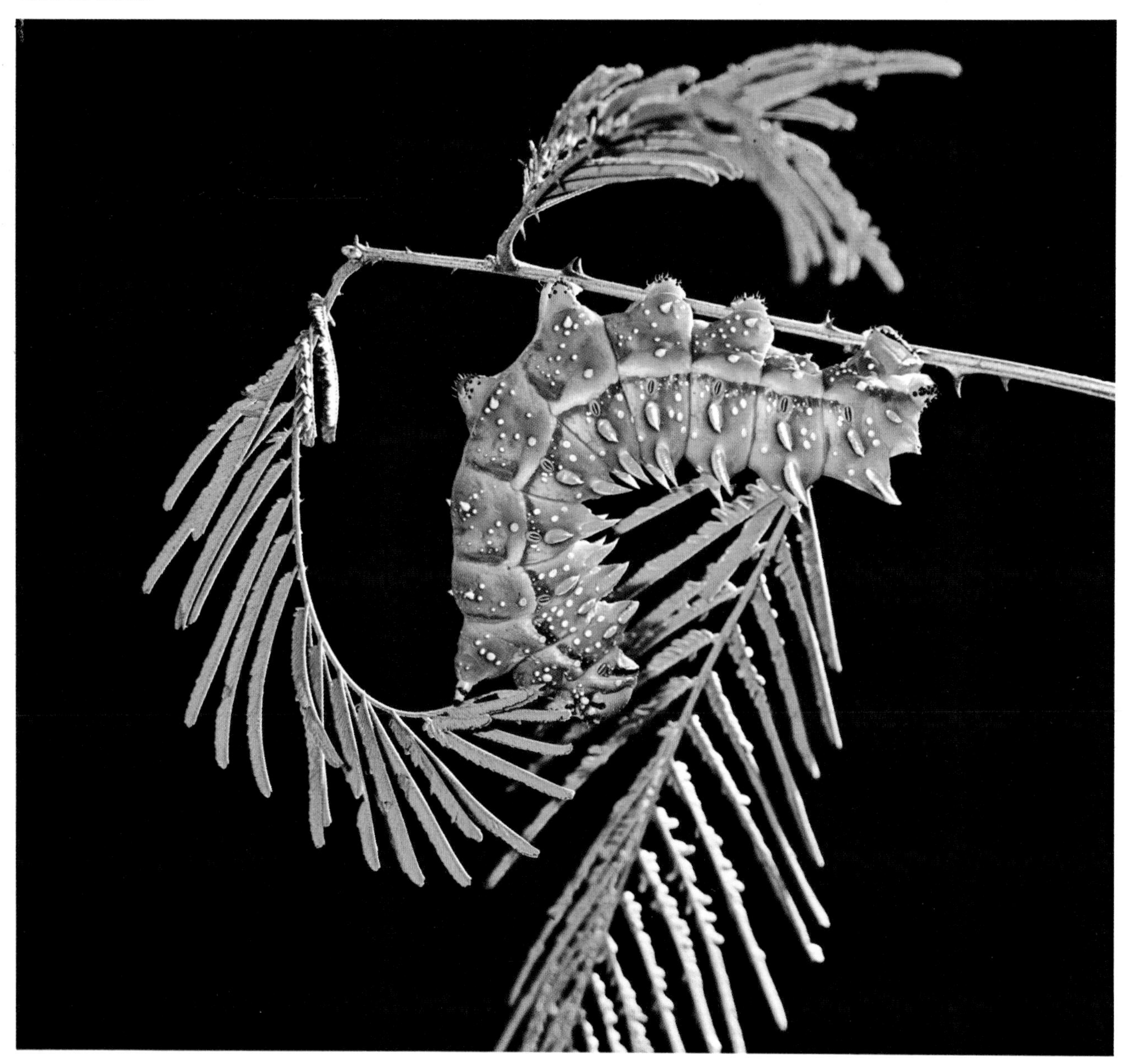

The tower walls are plaster-hard and solid, made of soil and small pieces of grass, bound together by special dry, hard termite excrement. Many are thought to be of great age. The main termite nests are in the ground beneath the towers. In the cavities and galleries of big termitaria there may be hundreds of thousands of termites, mostly workers and white nymphs in various stages of growth. The soldiers are quite amazing little automata with big shiny orange-brown heads. When alarmed they vibrate their abdomens from side to side against the tunnel walls, producing a tiny rattling noise, which is instantly taken up by other termites, the sound spreading out in a subterranean hiss.

Each termite nest also contains at least one pair of reproductives. The male is normal in size, but the queen has a hugely enlarged abdomen, very soft and delicate to the touch and in constant rhythmic peristaltic movement. She is a huge egg-producing machine, laying eight to ten thousand eggs a day at the rate of one every few seconds. The eggs are tended by the workers, which also feed and clean the queen and caress her with their palps. The workers feed and care for the developing nymphs, and feed the soldiers, which, with their enormous jaws, are incapable of feeding themselves.

With such a colossal rate of reproduction, termites are able to serve as food for a number of other animals without any diminution in their own populations. At certain times of the year, when the winged reproductives leave the nests in vast swarms, all kinds of animals gorge themselves on the brief abundance. Woodpeckers, wrynecks and other birds take a fair number of workers every day when

Termite towers are a conspicuous part of the African landscape, and the millions of termites that built them are of immense importance in the ecology of all lower lying tropical regions, wherever there are trees. Many animals benefit directly from the works of termites: dwarf mongooses, for example, use the towers as lookouts and refuges and the underground galleries as subways.

Above, right: Termites are most active above ground when the atmosphere is humid. After rain the workers get busy repairing and extending existing towers and constructing new ones.

Termites are not normally active in the open by day, but a light drizzle has enabled this work-party to forage without fear of desiccation. The workers drag dry grass and petals back to the nest while soldiers stand guard.

Hedgehogs are almost entirely nocturnal. Small, slow-moving terrestrial and nocturnal insects such as termites form a large part of their diet. In spite of their prickles hedgehogs are taken occasionally by Verreaux's eagle, owls and other predators.

Elephant shrews feed very largely on termites. This species lives in rather barren stony areas with scrubby bush cover. It bounds at great speed across open spaces between rocks and bushes; its runs are a series of widely-spaced scuffed patches rather than continuous paths.

they are poking about in dead trees for beetle grubs and other insects. Pangolins take termites, although ants are their main food. Large predatory ants known as stink ants, hunt termites individually. The smaller, equally fierce, ponerine ants go out hunting in an organized manner, in columns of up to 300; they return from their raids still in column, each ant carrying a termite in its jaws.

Termites are normally only active under cover by day, inside wood or underground, or in the little arcades of cemented earth they construct along the ground, up the outsides of trees or covering dead twigs. Here they can move about in the necessary conditions of darkness, humidity and coolness. But when the air is especially humid—immediately after rain or during a gentle shower—they are able to come out into the open to work either at rebuilding their towers or gathering food. After dark they can forage openly above ground, and are then very accessible to nocturnal small mammals such as shrews, hedgehogs and insect-eating rodents, whose droppings are often packed with the characteristic jaws of the worker termites.

Elephant shrews, though mainly diurnal, live chiefly on termites. They have a phenomenally acute sense of smell. Almost the instant their mobile

Besides the many lesser animals that feed largely on termites, two bigger mammals – the ant bear and the aardwolf – are almost exclusively termite-eaters; the ground pangolin takes termites as well as ants.

snout passes over a termite, the concertina tongue whips out and in, and the insect has gone. Termites are lapped up like this at a great rate, but are not swallowed immediately; they are stored in the elephant shrew's cheek pouches, to be chewed up thoroughly later.

Termites also support two much larger mammals, both almost exclusively termite-eaters. The aardwolf, the shyest and most elusive of the two, looks like a small and elegant striped hyena, but its feeble jaws and teeth make it incapable of chewing meat. It also lacks powerful digging feet, so its feeding is limited to picking termites from the surface or out of soft soil or rotten wood. Nevertheless, its tacky tongue sweeps them up with such efficiency that it is able to consume some 40,000 in three hours of feeding.

A much more powerful animal is the aardvark or ant bear. It has strong digging claws and is a very active burrower, digging into the ground with amazing speed. Its burrows may be a complicated maze of galleries with twenty or thirty entrances, spread over a large area; or a simple shaft with only one opening. The ant bear digs out termite mounds and laps up the insects with its very long sticky tongue. In a night it may travel as much as 15 kilometres, going from one termite mound to another, and it rarely visits the same one on successive nights. Great holes around the bases of termitaria are characteristic of its work, and these holes, together with its abandoned burrows, are used as hideouts and nursery dens by a number of carnivores: spotted and striped hyenas, aardwolves, jackals and bat-eared foxes, and warthogs. Newborn warthog piglets are very sensitive to falls in temperature, and need a stable environment until they are a few weeks old. When the warthog parents have enlarged them and lined them with grass, ant bear burrows are ideal nurseries. Steinbuck too, may hide down ant bear burrows. Steinbuck are unusual small antelopes: when they are alarmed, they will often sink down into the grass and freeze until danger is past; or they may creep away and hide instead of running in the open. They

Two topi calves lay resting together on an old grass-grown termite mound while their mothers grazed nearby. When a cow joined them her calf stood up to suckle. From this vantage the mother can survey the surrounding countryside and watch for any approaching predator.

The male agama lizard is aggressively territorial. He perches on a boulder, bobbing his conspicuous orange head as a signal to other males in nearby territories. Subordinate males and females are brown and green, perhaps with orange spots on the flanks.

Spiny mice are gregarious and live among rocks. They feed on insects and succulent leaves, such as the fleshy *Notonia gregoriae* with its orange shaving-brush flowers. Their spiny coats may afford them some protection from snakes.

may also use ant bear warrens as nurseries and give birth underground. Even bats may hang up in them to roost, while ground squirrels, mongooses, porcupines, all sorts of small mammals, occasionally use them to hide or sleep in.

Almost as well tenanted as the big ant bear burrows are old termite mounds. These provide smaller gauge burrows for animals that find the baked earth of the bush too hard to dig in for themselves. D'Arnaud's barbets, which can dig in soft earth, sometimes nest in the ready-made hole provided by a termite ventilation shaft. Plated lizards, snakes, ground squirrels, dwarf mongooses, rats and gerbils are small enough to use the larger termite tunnels as bolt holes. Packs of mongooses may travel hundreds of metres in the safety of the subterranean galleries, and both the dwarf and banded species use old termite mounds as sleeping quarters at night, and by day as combined basking areas and lookout towers. Big antelopes too, stand sentinel on old or disused termite mounds, converted by such use into flat-topped grassy hillocks. Nile monitors lay their eggs in termite mounds after tearing open a rain-softened wall: the termites repair the breach, sealing the eggs into a predator-proof incubator. Even elephants make use of termite mounds, as scratching posts. A favourite tower may be worn smooth and rounded by their repeated rubbing and scratching.

Insulbergs In some of the most rugged parts of East Africa there occur huge isolated granite boulders known as insulbergs. These rise up out of the bush from plinths of jumbled rocks, their sides sheer and smooth, bare of any vegetation. Insulbergs are found in some of the oldest parts of the Earth, in parts of Canada and Australia as well as in Africa.

The jumble of rocks around the bases of insulbergs are inhabited by the same animals as are found in rocky places elsewhere, such as rock hyraxes, klipspringers and agama lizards. In addition, the rocks may also shelter colonies of spiny mice. Among grey rocks, spiny mice are grey; elsewhere they are usually sandy coloured. Their backs are covered with a dense coat of short, sharp prickles, perhaps a defence against snakes. The mice eat seeds and insects and the leaves of succulents, and like hyraxes, bask out on the rocks in heaps in the early morning sun.

One bird in particular is found in association with insulbergs. The freckled nightjar is a rather large, dark-plumaged species which perches by day among the rocks, well concealed by its superb camouflage. It nests directly on a flat rock, in spite of the great heat of the surface under the sun (page 102). When incubating it sits so tight it is possible almost to tread on it before it flies up. With its eyes half closed, it looks like a carcass of a bird baked dry on the hot rock.

Oases Throughout most of the wilderness water is very scarce, except immediately after rain. But in a few places water is forced to the surface by impervious underlying rock, and bubbles up in springs. There are many such springs throughout the East African bush. Some are boiling, like the springs at Lake Hannington; some may be only slightly alkaline and form pools of warm, translucent, bluish water, oases in the middle of desert bush. One of the largest is at Mzima in the Tsavo National Park. It is crystal clear and the outflow is so great that it is the main source of the permanent Tsavo River. Other springs with lesser flows soon lose themselves again in the sand. Some are large enough for fish, and even hippopotamuses and crocodiles, to live in; the smaller ones have a rich fauna of tadpoles, water-boatmen, whirligig beetles and dragonflies.

In many places in otherwise quite arid country beautiful springs well up among the rocks, oases for thirsty animals. The water is often hot and slightly alkaline—hence the blueness. Milkweed plants, on which monarch butterflies lay their eggs, have gained a root-hold in rock crevices here.

Dragonflies abound in the vicinity of springs. They perch on grass stems overhanging the water, poised ready to swoop on any small flying insect. Their aquatic nymphs feed mainly on tadpoles.

The broad-banded rana spawns below the spring, where water spreads in a shallow pool before soaking once more into the sand. Its tadpoles are preyed on by dragonfly nymphs, while the adult dragonfly may in turn be snapped up by a frog.

The Plains

The transition from bush to plains is often abrupt, thickets opening up and giving way to rolling grassland within a kilometre or so. In a few places in East Africa the grasslands still remain in their original pristine state, vast green or golden plateaux full of great herds of grazing game and their attendant carnivores.

Grass is the most obvious plant of the plains. The dominant species in many areas is red oat grass: when it flowers it turns the plains into a russet sea. Another sweet and nutritious species is star-grass; *Pennisetum*, on the other hand, is generally tough and unpalatable. On alkaline flats a soda-tolerant grass flourishes, while on black cotton clay, which readily waterlogs after rain and cracks when dry, there are other nutritious species. In addition, there may be many small clovers and other legumes, as well as all sorts of small herbs which are sought out by the different animals.

The smaller plains game The fauna of the plains is predominantly grass-eating. The great herds of ungulates are the most spectacular, but there are very many smaller grass-eaters, too. Among these some of the smallest, but of immense ecological importance, are the harvester ants. They are large reddish-black ants which live in underground colonies of several thousands of individuals. Each year a colony collects millions of grass-blades and grass seeds, yet the ants do not normally have any adverse effect on the vegetation. In the dry season a circular area around a nest may be quite bare of plants except for the midden of husks piled up. Yet with the first shower of the rainy season, grasses and annual seedlings sprout on this bare patch. The ants are of considerable value in aerating, mixing and fertilizing the soil in places where earthworms are few or non-existent.

As ubiquitous as grass are grasshoppers. Wherever grass grows, from sea level to the snow-line, there are probably grasshoppers to eat it. They steadily consume their surroundings with mouthparts beautifully adapted for grinding up tough grass-blades. They are the second stage of a great food pyramid, for almost all insectivorous animals catch grasshoppers when they can, and some birds live almost wholly on them. However, grasshoppers are virtually invisible among grass. When feeding on moist greenery they are coloured bright green and when the grass bleaches they adapt their colour

Grass is the most obvious plant of the plains. There are very many species, some tough, some sweet and nutritious, but all utilized at different stages of growth by different grazers. The young reedbuck is feeding on the leaves of flowering Rhodes grass.

A young Thomson's gazelle buck, with its fine pointed muzzle, can delicately pluck small pods. Many nutritious herbs flourish among the grasses in wild pasture and supplement the grazer's diet.

A harvester ant takes a piece of grass down to its underground nest. Every year a colony of these ants collects millions of grass-blades and seeds, yet does not damage natural grassland.

A grasshopper of lush highland pasture chops through a grass-blade, then feeds it into its mouth with its forelegs. Among green grass, grasshoppers are green grass-coloured.

Cattle egrets following kongoni in the early morning. By using large herbivores as beaters, cattle egrets actually take half as much food again—and two-thirds as many steps—as when they forage on their own.

The springhare is one of the larger rodents of grassland. By day it stays underground in the large warrens it digs; by night it comes out to bound along kangaroo-like and to feed on grasses.

individually to white or yellow among dry grass or to brown, orange or black to match the soil. Some grasshopper-eaters, therefore, use other animals to find the grasshoppers for them: cattle egrets and wattled starlings follow large herbivores, and carmine bee-eaters ride on kori bustards or Abdim's storks, swooping off to snap up the flying insects disturbed by their feet. Many grasshoppers flash brilliant scarlet, yellow or blue underwings in flight, only to become invisible again the moment they settle.

Rodents are often very common in grassland, and there are several species that live on grass. The African hare and the springhare are the largest, and in north-western Uganda the Bunyoro rabbit is plentiful. There are many rat species. Grass rats live among grass almost anywhere, in very arid country or really swampy grassland, from sea level to 2,100 metres. They live in colonies and can be seen out during the day, scampering along runways in the grass, occasionally standing up on their hind legs like a ground squirrel, to get a better view. The grass rat can subsist on grass, but needs a supplementary diet of seeds to breed. The vlei rat lives entirely on grass; it has a special large caecum like a rabbit or hare, and vole-like laminated molars. With its very broad incisors it can chop down tall *Pennisetum* stems and split them to get at the soft vascular part inside. The runs it makes are most distinctive, wide and well-trodden and swept clean as the rat tramps along them on short legs, its long fur brushing the ground.

Rodents are the mainstay of the small mammal and bird predators, as grasshoppers are the mainstay of the insect-eaters. All sorts of medium-sized

Grass rats in typical tripod pose among Kikuyu grass. Many species of rats live in grassland, but this is the best known; it is common on farmland where it damages pasture and cereal crops.

Left: A superb glossy starling with a small grasshopper among flowering aloes. Grasshoppers are an abundant food and form the second stage in one of the great food pyramids. Almost all insectivorous animals catch grasshoppers whenever they can.

The vlei rat is *the* grass-eating rat. It chops down tough stems and slits them for the soft inner parts, leaving characteristic little heaps of shredded grass. Here it is feeding on red oat grass, the flowering heads arching above it.

The long-crested hawk eagle perches in a small dead feverthorn, surveying the grass below. It lives almost exclusively on rats, which it pounces on from its perch.

Africa is exceptionally rich in birds of prey of all sizes. The bataleur eagle is one of the most remarkable. It spends nearly all day on the wing and covers two or three hundred miles, scanning the ground for birds, rats, snakes and carrion. Two hours before dusk it perches again, its colours brilliant in the low sunlight.

to large birds of prey take rats and hares, and two smaller raptors, the black-shouldered kite and the long-crested hawk-eagle, live almost exclusively on rats. Mammalian predators—jackals, bat-eared foxes, wild cats, genets, even the tiny, black and white, weasel-like zorilla—take rats whenever they can. Some predators are diurnal, some nocturnal. Between them they feed on almost any creature that moves at any time on the plains, from small antelopes down to winged termites. The secretary bird, for instance, preys by day on all creeping things in the grass: frogs, locusts, small tortoises, game birds, lizards, snakes, as well as rats. When it spots something in the grass, its crest of long feathers is erected, breaking up the outline of its head and looking in silhouette from below like grass heads moving with the breeze. The puff adder, too, takes rats. It feeds mainly at night. Motionless, it waits on a run for a rat to come along, then lashes out with a lightning strike. The rat dies within a minute.

The big game of the plains Grazing insects and other small animals are almost as abundant as the grass itself, but from sheer size the big ungulates are the dominant grass-eaters of the plains. In parts of East Africa there still occur spectacular concentrations of big animals such as no longer exist anywhere else in the world. The most famous of the concentrations is found within the Serengeti National Park in north-western Tanzania and in the adjoining Mara Game Reserve in Kenya. Here over a million head of game live in 2,000 square kilometres of undulating plains interspersed with acacia woodland and riverine forest. There are many thousands of wildebeests, zebras, topis and gazelles, but also numerous giraffes, black rhinoceroses, elands and other antelopes large and small of over twenty species. Even bigger herds are found in the high rainfall grasslands of Uganda, in the Murchison Falls and the Queen Elizabeth National Parks. On both sides of the Nile live huge herds of big game, elephants, buffaloes and hippopotamuses, and enormous numbers of medium-sized antelopes. There are fewer species here than in the drier grasslands of Kenya and Tanzania, but the heavyweights occur in such numbers that their biomass—the weight of living animals to a given area—is probably greater than anywhere else in the world, and twice that of the Serengeti.

The grasslands of East Africa are able to support this great variety and profusion of large herbivores because almost all the species are separated ecologically, some by their food requirements, some by

Bat-eared foxes setting out across the plain for their evening's hunting. Small to medium-sized predators such as these are opportunists, taking rats whenever they can but snapping up other small game such as grasshoppers, lizards, young birds or whatever they chance across.

The secretary bird, an aberrant terrestrial eagle, stalks about the plains searching out any creeping thing for prey. It roosts on the flat crown of a thorntree, waking and stretching as the sun rises.

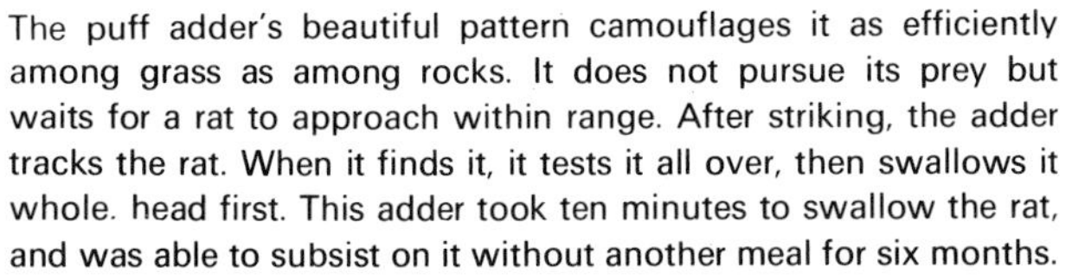

The puff adder's beautiful pattern camouflages it as efficiently among grass as among rocks. It does not pursue its prey but waits for a rat to approach within range. After striking, the adder tracks the rat. When it finds it, it tests it all over, then swallows it whole. head first. This adder took ten minutes to swallow the rat, and was able to subsist on it without another meal for six months.

Burchell's zebras, wildebeest and topi graze together. Although eating the same plants, these three species do not compete but complement each other in their use of the grass. While most of the herd feed heads down, several wildebeests stand sentinel on an old termite mound, each animal facing in a different direction for maximum coverage of the view.

their preferences for different types of grassland—swampy or dry, wooded or open. Waterbuck and reedbuck are found mainly in swampy grassland, and feed chiefly on grass. Impala and Grant's gazelles eat grass, but as they also browse the leaves of bushes, they are found in wooded grassland or bush.

Zebras, wildebeests, topis and Thomson's gazelles are grazers, so are mainly animals of the open plains. These four grazing species may feed off the same grass plant, but as each uses it at a different stage in the plant's growth, they do not compete. First zebras eat the outer part of the stem, which is too tough and unnutritious for antelopes but acceptable to zebras: they have incisors in both jaws for cutting through wiry stems and a gut organized to cope with a high throughput rate of low quality feed. Next topis, with their pointed muzzles, can get at the lower parts of the stems, while wildebeests' square-ended muzzles enable them to pick the horizontal leaves. Several days after this clipping the grass sprouts from the base, and Thomson's gazelles, with their small muzzles, nibble the new growth.

Buffaloes are the dominant large grazing animal in many areas. They feed mainly at night; before dawn they begin to move towards their daytime resting place, grazing as they go. During the day they lie up, usually in thick cover, then emerge again at dusk to begin feeding.

Buffalo grazing maintains a pasture in good condition. Buffaloes move along in a fairly compact group, munching mouthfuls of grass as they go, selecting not for species, but for stages of growth. They nose beneath the tougher top growth and bite off the lusher green shoots underneath. The tough, uneaten stems are trampled and act as a mulch. Next time the herd grazes that way it will find new growth stimulated by the previous cropping and easier to reach because of the trampling. Each time the pasture is used—providing it is not over-used—the grazing is improved. Patches of long grass which the buffaloes miss, either because it is unpalatable

A buffalo feeding at dawn on tall grasses at the edge of a swamp. Grazing by buffalo converts areas of tall unpalatable grass to a succulent greensward which smaller herbivores can utilize. A red-billed oxpecker is perched on the buffalo's shoulder and the birds in the foreground are blacksmith's plovers.

or has snags of fallen branches among it, serve to hide smaller herbivores; and in times of drought act as a useful reserve of standing hay which can be eaten by hungry animals.

Topi are also beneficial grazers because they can eat the drier grass stalks. Elephants and hippopotamuses are beneficial when they are not too numerous for they can reduce extensive areas of long, tough grasses to a short-grass sward which other animals can utilize. On a smaller scale, colonies of vlei rats improve a pasture by chopping down old stems of tussock grass. Cropping the grass stimulates its growth, so that full, even heavy grazing by a variety of herbivores keeps a pasture in an early, therefore succulent and nutritious, stage of growth. The grass is able to survive heavy mowing because its growth point is at the base not at the tip as in most other leaves.

The effects of normal grazing are subtle and often difficult to see. Only when over-grazing occurs, as when hippopotamuses become too numerous, are the effects obvious. An adult hippo eats 180 kilos of grass each night, so when at the beginning of the century, the numbers of hippos in Uganda increased enormously, the effect on the vegetation was drastic. Areas that were once grassland were grazed bare. The loss of grass prevented the spread of fires which keep down thornbush scrub, so this was able to regenerate at the expense of the grass, greatly reducing grazing not only for the hippopotamuses but for all the other grass-eaters as well. Since hippos are powerful animals with virtually no natural predators to keep them in check, the only means of preventing further pasture deterioration is extensive controlled cropping by shooting.

Cattle Under normal conditions ten or more species of large ungulates can live at high density over a wide area because they make use of the full spectrum of grasses, herbs, shrubs and trees. Their numbers are in delicate balance with the amount of plant growth, and even when food and water are short, over-grazing with consequent lasting damage to the pasture rarely occurs. In contrast, replacing the indigenous ungulates with equal, or even considerably lesser numbers of exotic species, as man has done over vast areas of Africa, may cause rapid and irreversible changes. The three types of domestic stock, cattle, sheep and goats, are changing the face of Africa as surely as they have already changed much of the Mediterranean region from lush forests and pastures into barren, stony hillsides.

Cattle select the sweeter grasses and feed by gathering the grass into their mouths and cropping it off almost to the ground. They take in the stems which their digestive systems cannot utilize, but which could be used efficiently by the better adapted zebra and topi. Sweet grasses draw their moisture from roots a few centimetres below the surface, and when the grass cover is removed by cattle during the dry season, this moisture evaporates rapidly; under very heavy grazing and trampling, the nutritious grasses are killed. Sheep are then brought in to eat the coarser grasses and after them goats, to tackle what the cows and sheep will not eat. Goats can eat almost anything; they pull up plants to get at the roots and climb trees to strip the foliage. They effectively prevent all regeneration of grasses and palatable herbs. At the same time tough, poisonous or spiny plants get the upper hand through lack of competition from other plants. The herds on their

A Tugen herdswoman carries a new-born kid as her goats graze towards the lake. Around the shore flat areas are still grass-covered, but the hills behind have been grazed and eroded to bare stony ground.

Over-grazing by goats has reduced this area of bush to bare stones in which only spiky and unpalatable plants persist—aloes, a patch of grey-green *Caralluma russelliana*, bayonet aloes and acacias. It is now the rainy season, yet no grass or seedlings sprout. How the hinged tortoise survives is a mystery.

Cattle assist erosion in various ways—here rather unusually by eating mineral-rich earth from the sides of a shallow ravine. There is little enough grass left in the area for them to eat, as the bare earth above the cliffs shows.

way to water follow the same routes each day, and their trampling hooves wear gullies in the soil down which water rushes in the rains. Some gorges in East Africa, now over 10 metres deep, started as cattle trails.

Pastoralism is a way of life forced upon many peoples by the climate of the areas in which they live. Rainfall is often less than 75 centimetres a year, and this is too dry for most crops. At best it produces marginal grazing for cattle, yet here the pastoral peoples of East Africa, the Masai, Samburu, Suk, Watutsi and so on, have traditionally close-herded cattle, sheep and goats for subsistence. These peoples do not eat meat, but their live animals provide them with a staple diet of curdled milk and blood, and with many other necessities such as dung for hut building and for fuel. The cattle are not usually slaughtered until they are old.

The cattle-dependent races are splendid people, physical giants compared with the agricultural races who in the past were stunted by protein-deficiency diseases, to which their young children were especially vulnerable. But the problem for the pastoralists is that their livelihood depends on the quality of the grazing. Where grazing has deteriorated badly, only very small amounts of blood and milk can be taken from each animal. The poorer the grazing, the more head of cattle are needed to supply enough blood and milk for each person; and the more cattle there are the worse is the over-grazing, so that each animal can supply less and less. Where peoples are dependent on cattle and have no other means of subsistence, they face starvation as the grasslands are turned into deserts.

The effects of over-grazing are more obvious and look worse in arid than in wetter districts. Today, in many areas where twenty or thirty years ago lush grassland supported game from elephants down, the end point of over-grazing has been reached: stony red soil barely covered by sparse thornbushes and spiky succulents, cattle skeletal even in years of good rain, and the game virtually gone. Ants and termites are sometimes blamed for damage to pasture because their nests can be more easily seen and appear more abundant in over-grazed areas; yet grasses and trees survived their onslaught for millions of years until man and his cattle destroyed the balance. In the end, even the ants and the termites may die out from these areas.

Cattle have been in Africa for four or five thousand years but they have failed to adapt fully. They suffer from heat stress and need to drink at least once a day; and endemic diseases cause a heavy death toll. The most serious of these diseases is *trypanosomiasis*, or sleeping sickness, known as nagana in cattle. It is transmitted by the tsetse fly. The flies feed by piercing the skin and sucking blood until they are bloated. While doing so, single-celled protozoa—trypanosomes—are transmitted to the host in the saliva of the fly. These undergo part of their development in the blood of the host, part in the gut of the fly.

Tsetses are endemic to Africa south of the Sahara and there are over twenty species. Some live in thick forest, some in bush, some near lakes. Wild mammals, birds or reptiles, which evolved together with the tsetse over millions of years, do not succumb to the diseases which they carry, but introduced cattle do, and so are impossible to keep in a third of the continent, wherever tsetse flies occur.

In some parts of Africa from Uganda southwards, man has attempted to eliminate the tsetse from potential cattle areas by shooting out the game which acts as a reservoir for the trypanosomes, by felling trees and bush that harbour the insects, and by trapping the flies and spraying the countryside with insecticides from the air. Such operations cost millions of pounds and are only marginally successful. Despite the shooting of many millions of big animals, small and elusive game such as bush pigs and duikers escape the slaughter and continue to act as reservoir hosts for the protozoa. Dense thickets of thornscrub invade the cleared bush, so that the land, even if it is free of the tsetse, becomes

Part of a vast herd of Samburu cattle returning home to their village in the evening. The rainy season is well advanced yet the cattle remain pitifully thin. In a few years' time this trail may become a ravine down which flash floods briefly roar during the rains. It is not the pastoral system itself which damages the habitat, but keeping more cattle than the grass can sustain.

Three impala bucks of different ages pause to nibble grass as the bachelor herd crosses the dusty bed of a former lake. In contrast to the poor-looking cattle they are beautifully sleek; by feeding on a variety of leaves and twigs as well as grasses they are able to utilize an arid habitat to the full. In this tsetse area they are probably infected with trypanosomes, but unlike cattle they will not succumb to the disease they are carrying.

useless for grazing cattle. The policy, so terribly costly in natural resources, was a failure and has been suspended. Nowadays, the tsetse is looked upon by conservationists as the saviour of Africa, conserving the soil from over-grazing and erosion. At the same time the environment and the wild ungulates in it are acknowledged as a unique natural resource, to be utilized wisely, as well as enjoyed for the pure pleasure of seeing beautiful animals in the superb unspoilt African countryside.

The great beasts of prey The big herbivores are in constant fine balance with the plants of the environment. In just as fine a balance with the herbivores are the great carnivores. The chief predator of plains game is the lion, and its prey is generally the dominant herbivore on its home range. In some places this may be buffalo, in others wildebeest or zebra; it does not show a particular food preference, but usually takes what it can get most easily. A lion can hunt by day but more often does so by night. It employs various techniques, sometimes hunting singly, sometimes in groups. In thick cover a lion will stalk its prey by stealth, creeping belly to the ground, freezing when the antelope's head comes up alert, moving only when the antelope lowers its head again to feed. In the open it will move purposefully after its prey, quickening its pace as the animal turns to flee; then with a powerful spurt of speed it gains on the antelope and bowls it over. Sometimes lions lie up and wait in dense cover, often near a waterhole among bushes which the game must pass when they come to drink. At times a lion will show itself or roar or give its scent on purpose to stampede the prey towards other lions waiting in ambush. A pride usually kills a large animal every three or four days, and may kill a hundred in a year.

When seriously alarmed, plains game flee at full speed, high jumping in every direction. Impalas, particularly, do most spectacular leaps. This serves to confuse any predator whose hunting technique requires concentration on a single individual, and also enables the fleeing antelopes to keep watching the enemy as they bound through bush cover. The black and white markings on the sides and rumps of many antelopes, and the stripes of zebras, accentuate the dazzle effect as the animals streak about.

The largest purely diurnal predator is the cheetah. A sprinter, and the fastest animal on land, it fetches down its prey by sheer speed. It may take any one of twenty-five different prey species, but on the plains Thomson's gazelles are its main food, while in bushy areas it takes chiefly impalas. Like a lion, a cheetah selects a particular animal before trying to catch it. It may stalk its prey for up to half an hour, with head and body low, taking advantage of every termite hill or bush to cover its approach. When it is close enough for the final rush, the cheetah dashes out, dodging and zigzagging after its fleeing prey until it can bowl it over and grab it by the throat.

The third of the great African cats, the leopard, is the most secretive of the three. It is usually found in forest or wooded places, and takes its prey by leaping on it from a branch or from its ambush in thick cover. In the forest it feeds mainly on duikers and other small antelopes; up mountains it lives on hyraxes; among rocky hills and cliffs it takes baboons and klipspringers. It also preys heavily on the newborn young of plains game.

It has been found that leopards more often kill male animals than females, except in the breeding season when the very young are taken indiscriminately. The same is true of lions and cheetahs: their prey is more often males in their prime than females or yearlings. Among plains game, adult males make up a fifth of the population, yet of kills made by the big cats, three-fifths are adult males. This is because males are easiest to catch, being more likely to be separated from the herd. In many antelopes a lone individual is almost invariably a subordinate male who feeds by himself while the females and young stay together. In others, such as kob and Thomson's gazelle, the solitary males are the successful territory owners. Bachelor Grevy's zebras are solitary and territorial, while the family herds graze in separate areas. Animals in herds rely on many pairs of eyes, ears and nostrils to detect danger and so may escape a predator whereas a lone male is more readily taken unawares.

The effect of predation is therefore to reduce the number of males. In a harem society, however,

A leopard draped along a branch at sunset with its tail hanging down like a bell-pull. It will stay feeding through the night on its kill, a waterbuck calf, which it has dragged up the tree out of the reach of vultures, jackals and hyenas.

A bachelor Grevy's zebra patrols his territory in the early morning; mist and cloud are low on mountains of the Matthew's range. Solitary adult male herbivores are most vulnerable to predation by the big cats, and make up over half of their kills. When this picture was taken, the stallion on the adjoining territory had just been killed by lions (below).

A young lioness, her belly distended with meat, half-heartedly gnaws at the remains of the Grevy zebra stallion. Her pride made the kill before dawn; now it is evening and they have been feeding all day. The other lions lie gorged and panting nearby.

only a few males are needed to maintain the population. The excess males promote healthy competition and ensure that only the most virile sire the next generation. They also provide food for the carnivores, who thus keep the herbivore populations stable. Plains game are capable of enormous increases in their populations if unchecked. Although they give birth to only one young a year, they can multiply at a rate which can double the population every four years. If they are not preyed upon, they soon outgraze themselves, so predation is in fact an essential process in preventing overpopulation. The hundred herbivores a pride of lions kills in a year represents only the annual breeding increment and under normal predation, in normal climatic conditions, the herbivore populations remain remarkably constant.

But what controls the numbers of the predators? To some extent, especially among lions, territorial behaviour may be important in limiting numbers. Unlike the ritualized combats between rival antelope males, which are usually resolved without serious injury to either contestant, war between territorial male lions is often a fight to the death. Also, predators breed at a rate that keeps pace with the food supply. If the plains game decline, mortality among lion cubs may be high, as many as half

The cheetah is the fastest animal on land and preys on gazelles and antelopes. It does not hunt for itself alone, but leaves much of its kill for the scavengers. Replete, it suns itself, idly scratching.

Hunting dogs, the most efficient killers of the plains, are gregarious and hunt in packs in an organized way. They play an important part in the natural balance, controlling the numbers of herbivores like any other predator; and like the big cats they can seem lethargic in the heat of the day.

The web of life in grassland. Grasshoppers, grass-eating rodents, and plains game large and small are the bases of the three great food pyramids of the plains. In complex natural communities very few food chains are simple; very few animals avoid falling prey to other animals. Most grass-eaters are preyed on by either insect-eaters or by carnivores; carnivores may eat the insect-eaters and the carnivores may in turn be eaten by larger carnivores. Only the great super-predators and scavengers at the summit of the pyramid are free from the threat of predation. This diagram does not attempt to show all the animals involved in the intricate grassland pyramid, but illustrates representatives of the various groups.

THORPE

a litter being lost in the first year. But in places where game is abundant, mortality of cubs is very low. If the buffaloes, zebras and antelopes increase, then lions rear more young and therefore need to kill more prey. Thus an equilibrium is maintained.

However, the delicate balance between herbivores and carnivores is never simple, especially in an environment so rich in species of both predators and their prey. On the plains the situation is complicated by the presence of two other big predators, hunting dogs and spotted hyenas; all five major predators may prey on the same big grazers.

The hunting technique of hunting dogs is quite different from that of any of the cats. Hunting dogs are long distance runners, with impressive staying powers. They are not built for great speed, yet have sufficient to run down any antelope. They hunt in packs, and after their leader has selected the prey, they follow it until it begins to tire. The lead dog then increases speed, catches up with the animal and further exhausts it by repeatedly snapping chunks from its belly or from any part it can grab. Finally the animal either falls exhausted, or several dogs bring it down and rip it to pieces. Small antelopes are killed very quickly, but larger beasts take longer to die, though their death is not necessarily any more lingering than if it were inflicted by a lion or a cheetah. In fact, hunting dogs are the most efficient killers of all the large predators. Once the prey is selected, it does not escape. The majority of antelopes killed by hunting dogs are, again, territorial males, and because the dogs are nomadic, the herds are never harassed unduly, or over-cropped by them.

Hunting dogs kill by day; spotted hyenas by night. Hyenas are not merely scavengers, but are also major predators in their own right. After dark a pack will set off in a seemingly leisurely way until prey is sighted. They begin to harass the herd until

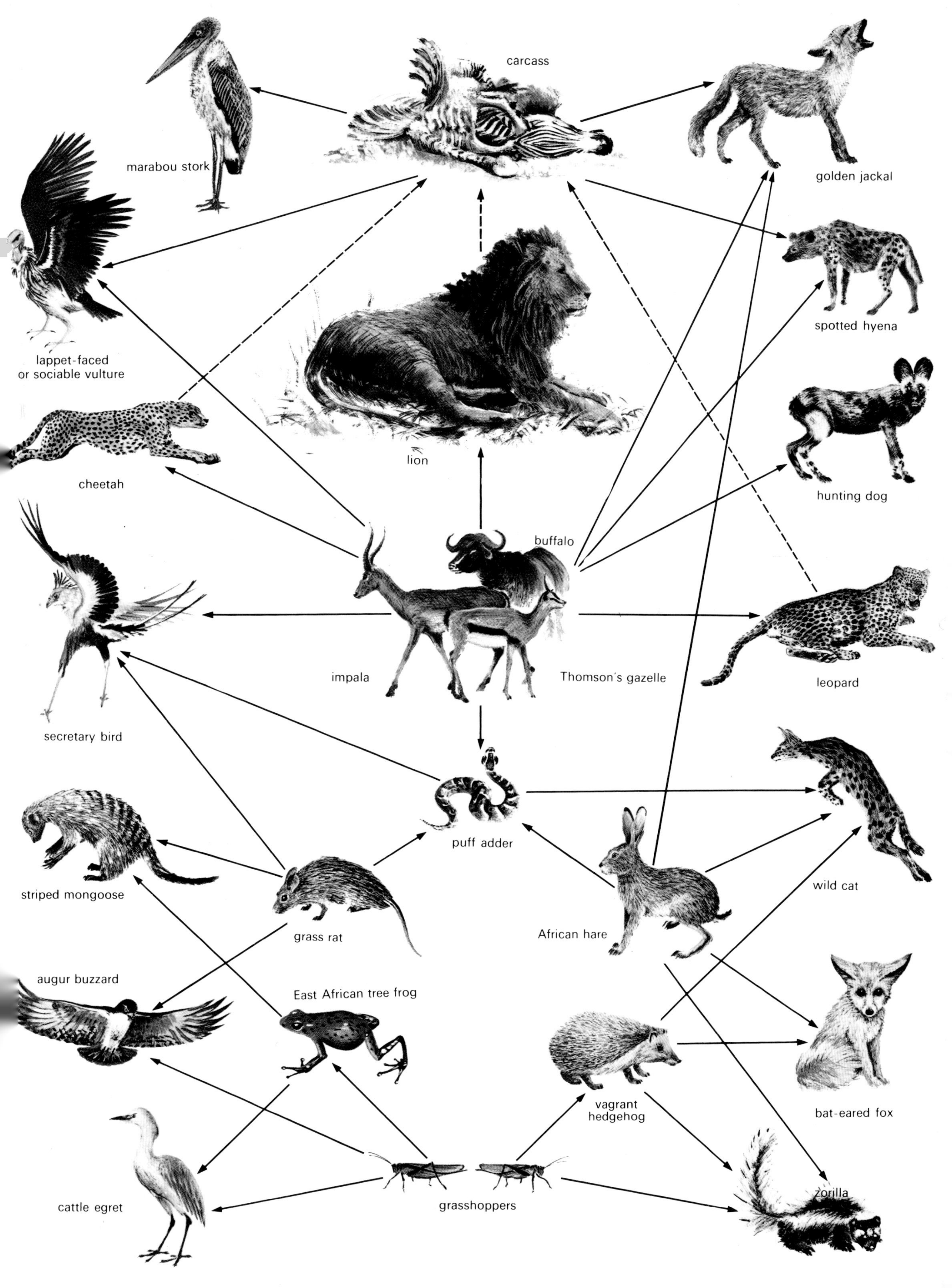
carcass
marabou stork
golden jackal
lappet-faced
or sociable vulture
spotted hyena
lion
cheetah
hunting dog
buffalo
impala
Thomson's gazelle
leopard
secretary bird
puff adder
striped mongoose
wild cat
grass rat
African hare
augur buzzard
East African tree frog
vagrant
hedgehog
bat-eared fox
cattle egret
grasshoppers
zorilla

The first vulture to arrive at the old lion kill at daybreak was this hooded vulture, lighter and more mobile than the griffons. When it began to feed, the bare skin of its head was quite pale. A little later griffons started to drop from the nearby trees and the hooded vulture blushed in defensive display.

they can single out a sick or very young animal. The whole pack then concentrates on this and quickly brings it down, tearing it to pieces. Very little is left of the carcass, for hyenas have the most powerful jaws in proportion to their size of any living animal.

All the major carnivores, apart from hyenas, may leave pickings from their kill. There is meat or offal or bones enough to feed a variety of scavenging animals as well as the predator that made the kill. Cheetahs eat only the meat of their prey, leaving all the rest—skin, bones, digestive tract and its contents—for the scavengers. Leopards take their kills up into trees but when they have finished eating, arboreal scavengers such as crows and ravens, even bataleur eagles, pull the remnants about and bring them to the ground. There hyenas, vultures, even ants and other insects, finish off the hide and bones. Where a lot of large animals have been killed, there may not be enough hyenas to dispose of all the bones. On the route of the wildebeest migration (page 117) many bones and horns remain scattered above ground. Some rodents such as rats and porcupines may gnaw at them, but in the end weather and bacteria break down the bones and specially adapted fungi grow on the horns.

Hyenas are the main scavengers by night, but during the day their place at the carcass is taken by vultures, which depend on lions to provide most of their food. There are ten species in Africa, six of which are common in East Africa and may even feed together around the same carcass. Although competing for food at times, the species are generally separated by their preferences for different types of country. Rüppell's griffon vulture is a bird of dry bush and desert; the white-backed vulture prefers grassland dotted with acacia trees. The former nests on cliffs; the latter in tall riverine trees. All vultures cover such vast distances in their search for food, it is scarcely surprising that several species sometimes converge on the same piece of carrion.

Vultures find their food mainly by soaring high in the thermals and watching not only the ground for lions or their prey, but also the behaviour of other vultures. Cruising effortlessly hundreds of metres above the ground they can spot another vulture dropping down a long way away. Other vultures even further off in turn see them descend, and so all soaring in a wide radius may converge when a kill is discovered. Animals on the ground also watch vultures descend, especially jackals and hyenas which sometimes locate food this way. Sometimes vultures perch in dead trees, keeping watch over lions, hoping they will rouse themselves and make a kill. Or the lions may already have a kill nearby, which they have dragged into the shade and are guarding. Vultures often have a fruitless vigil, driven off by lions, jackals, hyenas, even marabous, whose long reach and huge sharp bills give them an advantage. When vultures alone are at a carcass, the hungriest has precedence. Any famished new arrival puts on a great display at touchdown, bounding towards the carcass with neck arched and wings spread. The less hungry birds that have already bolted some meat give way before this apparition.

However unlovely vultures may appear, they are beautifully adapted to fill their role of scavengers. Their long sight and powerful wings often bring them to carrion that may not be discovered by terrestrial scavengers. Their function is therefore largely complementary to that of the jackals and hyenas.

The plains, by virtue of their openness, offer a unique opportunity to see at a glance the basic relationships of what is a complex ecological system. The herbivores can be seen consuming vast quantities of grass while the carnivores that keep their numbers in check lurk in sheltered places. Overhead, vultures wheel, while perhaps a solitary jackal trots purposefully along a well-worn trail. The overriding impression is one of harmony, with each animal perfectly adapted to fit a particular way of life.

Just after dawn, griffon vultures descended in a squabbling mass onto the old kill, driving off the small hooded vulture. Vultures are an essential part of the ecology of the plains, rapidly and efficiently disposing of carrion during the day.

The black-backed jackal feeds on any small game, and as a scavenger is also attracted to lion kills. This jackal is lying near the remnants of an old kill just before dawn. It is fully fed and is making way for the hungrier vultures; it will soon trot along home with its mate to their den in an old ant-bear burrow.

The Dry Season

A Thomson's gazelle ram on the parched plains seems to be making its way towards a shimmering lake. Doubtless it is not deceived by the mirage, since an antelope depends as much on its nose and its memory as on its sight to guide it to water.

Previous page: The skull of a greater flamingo lies on the cracking mud. During the dry season lake waters shrink rapidly; puddles, waterholes and even rivers dry up, leaving much of East Africa a waterless wilderness.

The African lungfish swimming normally, and below, in the cocoon in which it passes the dry season.

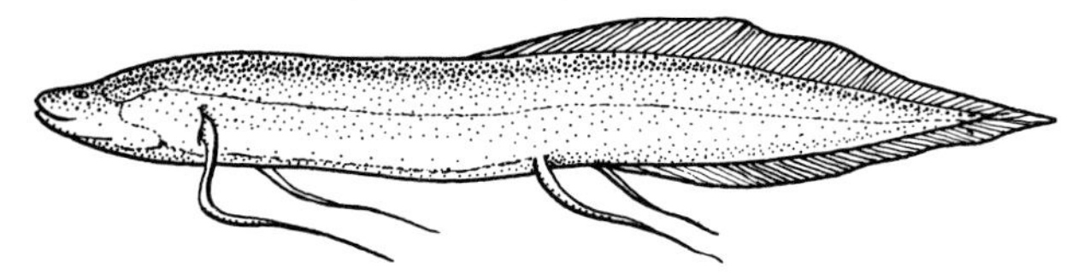

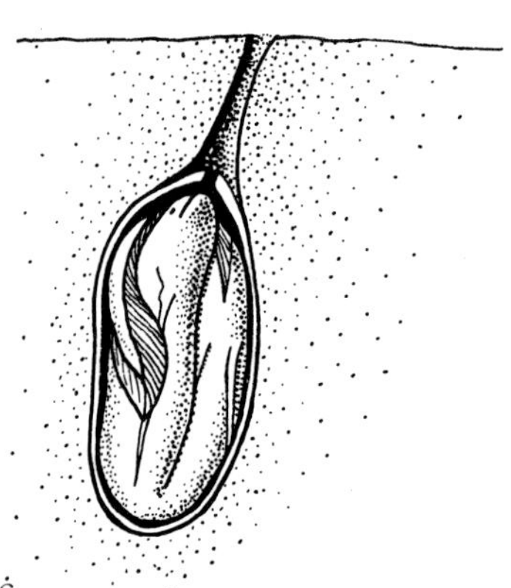

MILNE

East Africa is essentially a hot place. Even in the highlands, the dry season daytime temperature may be 27°C. On the plains, midday temperatures may exceed 38°C. During the dry season the sky is often cloudless throughout the day, or with only wisps of high, dappled clouds in the afternoon. When humidity is low, more solar heat gets through to the ground in the daytime, and more is lost from it during the night. So in the cloudless dry season, with the sun vertically overhead at midday, the heat may be intense, and the nights that follow correspondingly colder. In the heat of midday the scenery quivers and vibrates, and silvery mirages appear like lakes above bone-dry ground. Moving game in the distance seem to be wading through the shimmering water. The sizes of animals are distorted by heat haze and distance, so that gazelles may look like buffaloes, or a jackal like a lion. The sense of intense heat is accentuated by the faded colours of the landscape; the dead greys of leafless thorn bushes, the pale gold of dry grass, with blues and purples in the distance.

At this season there is often a dearth of small animals. Butterflies are few; grasshoppers, small spiders, aphids, even ants, scarce; caterpillars apparently non-existent. Toads and invertebrates

The small African catfish, a species with accessory air-breathing organs, can survive in damp mud when rivers dry out.

that are very vulnerable to desiccation, hide during the day in the relative coolness and humidity of the earth beneath logs and stones. Other animals go to ground for the duration of the dry season and pass it in summer sleep or *aestivation.* At the start of the dry season, these animals burrow into the ground before it becomes too hard. During aestivation, their metabolism is greatly lowered, as it is in animals that hibernate through very cold weather. Tortoises and terrapins aestivate, as do some frogs and very many invertebrates, and the fat mouse probably also does. Giant snails cement their shells shut with mucus which dries into a hard cap to reduce evaporation.

Among East African animals that aestivate, one of the most unusual is the lungfish. As soon as its swampy habitat begins to dry out, it buries itself in the mud, curls up and secretes a mucus which hardens as the mud dries, to form a cocoon in which the fish's body stays moist. Lungfish, as their name implies, possess primitive air-breathing organs which correspond to the lungs of higher vertebrates. No other living fishes possess true lungs, although some breathe air by other means, modifications of swimbladder, gills or gut. Lungfishes still have gills, though much reduced, with which they can breathe in water. While buried in the mud the lungfish is able to go on breathing through perforations in the top of the cocoon, and so exists until rain releases it.

The African catfish also buries itself in the mud during the dry season, as it is one of the fish with accessory air-breathing organs. In swamps and other poorly-oxygenated water it surfaces frequently to gulp a mouthful of air and instantly shoots down to the bottom again. It is extremely wary, and under non-drought conditions would not be caught napping at the surface by fish eagle or heron. When the pools begin to dry up it is quite comfortable breathing air and guzzling up the less fortunate smaller fishes stranded gasping and flipping in the shallows. Extremely voracious, it is capable of such an intake of food that its belly becomes spherical, but in shallow water it is just as likely to be taken itself by all sorts of opportunist fish-eaters. Indeed, in some districts fish eagles time their nesting to coincide with an abundance of catfish in dry season low water.

Water birds

Other large water birds besides the fish eagle nest in the dry season and hatch their young when lakes and rivers are drying up and aquatic prey of all kinds is easiest to catch. Herons, egrets, storks, spoonbills and ibises nest in big mixed colonies in trees, usually on an island or lake shore, sometimes far from water. Smaller birds that nest on the ground near water and so are vulnerable to flooding, often breed when the waters are receding. Gulls and terns, plovers, skimmers, pratincoles and waders nest on sandbars and shores. In the dry season, too, the receding waters of the soda lakes expose acres of wet mud which lesser flamingos need for breeding on.

Lesser flamingos breed chiefly on Lake Natron, in enormous colonies 12 kilometres out on an alkaline mudflat, hidden from the shore by mirage. Another breeding site in East Africa is a crater lake on a small volcanic island on Lake Rudolf. On Lakes Nakuru and Hannington the

The lesser flamingo nests at the beginning of the dry season when receding water exposes mud flats in ideal condition for nest-building.

Kittlitz's sand plover may nest on sandbanks or lake shores during the dry season to avoid the danger of floods. When the bird leaves its nest it does a little dance around it, scuffing sand over the eggs to hide them and to shield them from the hot sun.

A large crocodile basks on water-smoothed rocks beside the permanent Galana River, whose waters are becoming shallow now as the dry season advances.

Right: A pair of Masai giraffe resting and cudding at midday in the shade of feverthorn trees.

mudflats are not isolated from the shore, and though the flamingos build nests they do not breed successfully. The nests are pedestals of mud which the birds scoop up with their bills. On the baking mudflats the midday temperatures are high, but the long legs of adult flamingos carry them above the worst of it and on the top of their mound nests the temperature is only at blood heat. The disadvantages of high temperatures and glare are outweighed by the advantages of isolation from most predators.

In very hot areas such as Lake Rudolf crocodiles leave the water before sunrise and come out to bask. Almost all crocodiles come out onto the shore, except the big territorial males who stay in the water, lying half-submerged or patrolling in the shallows. When the sun becomes too fierce, the basking crocodiles retire to the shade or go back into the lake. By following a daily pattern of alternating periods in the sun, in the shade, and in the water, they are able to maintain a body temperature varying only a few degrees above or below 25°C. They also regulate their temperatures by basking with the mouth open, to radiate heat and evaporate moisture.

The heat of the day

During the hot dry season the majority of big game animals spend the heat of the day resting in the shade of bushes and trees to avoid overheating and dehydration. Elephants feed in the early morning and again in the evening as they make towards water, but from about mid-morning until mid-afternoon they remain quietly in the shade, standing beneath a thorntree or occasionally lying down. Black rhinoceroses, too, may stand beneath thorntrees in their characteristic dejected-looking, head down attitude, relying on their attendant tickbirds to alert them to approaching danger. Warthog males spend the greater part of the day lying up in the shade of bushes, but females and young must spend more of their lives feeding, so only lie up during the very hottest part of the day. The exuberant savannah chimpanzees fall quiet and doze through the heat of midday.

Most of the large herbivores rest in shade during the heat, usually standing, but sometimes lying down with head up while they chew the cud. Ruminants rarely lie on their sides as this position interferes with the smooth working of their complicated digestive systems. A few antelope, such as Bohor reedbuck, make form-like shelters in tall grass where they lie up like hares. Impala rest together among bushes, backs to the breeze, most of the herd lying down, with some older females standing to keep watch.

While herbivores are resting in the shade, they may shut their eyes but they do not often sleep. Their rhythmically cudding jaws, and the twitching of ears and tails, show they are only lightly asleep. Only occasionally does an animal go into a brief, deep sleep. A lightly sleeping antelope remains with its head up but, in contrast, one going off into a deep sleep stretches its neck along the ground and allows its head to flop over. Antelope do not sleep deeply for longer than four minutes at a time, but zebra may stretch out and sleep soundly for ten minutes. During the night zebras lie down together to sleep, with one or two standing sentinel. During the day, however, they do not sleep communally; an individual will lie down and sleep by himself, with zebras or other plains game standing watchful close by.

During the heat of the day the big cats appear most lethargic. Leopards loll along high branches of tall trees, lions collapse in the shade of bushes, especially near half-eaten prey, which they may continue to guard from scavengers. A lion that has

gorged itself often looks uncomfortably hot, not only at midday but in the cool of the evening as well. This is because he must pant heavily to lose water so as not to become overheated. Man and horses, including zebras, have well-developed sweat glands, but most other mammals, including lions, have few. Hunting dogs have sweat glands in the feet, so cool themselves by evaporation from the pads as well as by panting and dripping saliva from the tongue. The ears of some animals are secondary devices for radiating heat from the body: elephants constantly fan their great ears backwards and forwards, and antelopes flick their ears rapidly.

Birds fare rather better in a hot climate than do most mammals, since their body temperature is higher. Whether they inhabit the tropics or the arctic their temperature is 41°C. when inactive, and up to 43°C. during intense activity. Nevertheless they occasionally need to lose heat, and, having no sweat glands they, like many mammals, must pant rapidly to cool themselves. Some species pant at the rate of up to 300 respirations a minute, and like the lion, look as if they are in distress from the heat.

An example of the way birds can withstand high temperatures is the freckled nightjar (page 73) which lays its eggs on completely exposed bare rock. At midday the dark-coloured rock absorbs all the sun's heat and feels baking hot to the human touch; yet the incubating bird manages to maintain its eggs at the right temperature for development.

The animal that is best adapted to life in hot places is the camel, a true desert animal kept by tribes in the northern districts of Kenya but not brought much further south than the semi-arid country around Lake Baringo. The camel is able to lose heat by sweating, but need not do so except in extreme heat because its body temperature can rise several degrees without harmful effects.

Grant's gazelles and Beisa oryx also allow their body temperatures to rise by up to 6°C. during the day, from about 39°C. to 45°C., before they start sweating. Eland may allow a three-degree temperature rise. Such antelopes have a special system for cooling the brain more than the rest of the body, to avoid brain damage which would otherwise result from high body temperatures. Grant's gazelles are therefore able to remain in the open during the hottest weather, while most other animals seek the shade. They have very shiny coats of short hair which reflect the heat, as do

The camel is a true desert animal and looks a little incongruous among the comparative verdure around Lake Baringo. A large youngster follows its mother closely, holding onto her tail.

Left: A vulturine guineafowl makes its way to the waterhole at midday, feeding as it goes. Even birds used to living in hot dry bushland need to pant occasionally to cool themselves.

A rapidly panting small bird can look as if it is suffering extreme distress from the heat, but a large bird like the yellow-billed stork simply seems to be enjoying a joke. Lesser flamingos achieve some heat-loss all the time, paddling in the cooler lake.

The pale sandy coloration of the Grant's gazelle blends with the dry grass. It is one of the few mammals that habitually stays out in the open under the powerful midday sun.

A small herd of wildebeests grazing the parched yellow plains. The grass looks dead, but is still nutritious and will sprout afresh as soon as rain falls. In spite of their dark coloration, wildebeests do not suffer from heat-stress and do not need to seek shade except in the very hottest weather.

goats' glossy hard coats and the iridescent plumages of starlings. Their very pale sandy coloration may also contribute to heat regulation by reflecting much of the sun's radiation. On the other hand, the pale coloration may be mainly protective; Grant's gazelles exactly match in tone and colour the yellow plains grass in the dry season.

It was once argued that pale coloration was an adaptation to a hot environment because it reflected heat, while dark colours that absorb heat were thought to be an adaptation to a temperate climate. But domestic sheep in semi-desert places may be black, white or pied, or strikingly coloured with foreparts white and hindquarters black. Zebras, too, are black and white, reflecting and absorbing in almost equal amounts. (The zebra's stripes are, however, generally regarded as a protective device which renders the animal almost invisible at dawn or dusk or in moonlight; and as a dazzle-effect for confusing predators at close range.) Even dark-coloured wild animals such as wildebeests, buffaloes, rhinoceroses and elephants do not always seek shade at midday. Buffaloes spend most of the day resting and cudding in some peaceful place. Most often this will be in the densest cover they can find, but not because they need the shade. Their chief requirement is tranquillity, so that in places where buffaloes are not persecuted, their resting place, except in the very hottest weather, may well be in the open.

Some black rhinoceroses show a marked preference for dusty depressions and sand bowls as resting places, and they, too, will remain out in the full sun. A rhinoceros has to get up about every hour and a quarter while he is resting, to relieve the cramped sleeping posture for ten minutes or so,

Black rhinoceroses sometimes rest out in the full midday sun—here in the baking heat of a vast dustbowl whose fine, whitish alkaline soil was once part of a lake bed. The pair appear to be asleep, but are only dozing. A call of alarm from their attendant oxpecker would alert them instantly.

Warthogs at a wallow in the late afternoon. The smaller one is enjoying a muddy drink, while the larger one cools itself by rolling luxuriously, coating itself all over with wet mud. As well as cooling, the wet mud may act as a soothing plaster for minor sores and irritations, and a deterrent to biting flies and other parasites. When dry it is an effective camouflage among termite hills and dusty bush.

but it will remain in the same dust bowl, alternately lying down and standing up, for ten hours at a stretch. A resting rhinoceros' great hulk blends surprisingly well with the dull grey thornbush scrub, especially when both animal and vegetation are coated with a layer of the same dust. The small cloud of dust which the rhino raises as it exhales is often the only thing that gives away its position when it is in thick bush.

Living without rain

Leafless thornbushes offer little shade, so poorly-adapted animals must endure the heat as best they may until the evening brings coolness. Then they become active again, making their way towards water, which means not only a drink but a heat-dissipating wallow in the mud. Rhinoceroses, elephants, buffaloes, warthogs and spotted hyenas wallow as an efficient method of cooling the body and disposing of heat accumulated during the day. A wallowing animal lies in the mud to plaster each side of its body, and may even roll right over to plaster its back as well. Ostriches, which often

A domestic donkey and her foal are not greatly troubled by the poor grazing and lack of water during drought. They are descended from wild asses, true desert animals, and can withstand harsh conditions and an extreme climate.

Right: Impala come to water in a close herd at sundown. Some suck quickly from the edge, while a ewe remains watchful. Marabou storks rest, some on their heels, after drinking; they have fed well on nearby carrion.

During the dry season, game comes to water at any time of the day, but mostly in early morning or evening. Here wildebeest drink at sunrise as three impala bucks depart.

behave more like ungulates than birds, will sometimes wallow in water, submerging all but the neck and head. Zebras and antelopes do not wallow; perhaps this is a luxury enjoyed only by the largest or best protected animals. The majority of herbivores come to water only to quench their thirst, and do this as quickly as they can. Impala come in a close herd; each animal drinks from the very edge for less than half a minute, some remaining watchful while others drink. Then they and other plains game leave as quickly as they came, for at the waterhole they are especially vulnerable to all kinds of predators, from crocodiles and pythons in the water, to leopards and lions in the bushes.

In contrast to the hastily sucking antelopes and zebras, lions take their time at a waterhole. Their method of drinking is slow and laborious; it may take twenty minutes for them to get their fill by lapping. Lions prefer clean water from a flowing river, but they will drink from a stinking stagnant waterhole if necessary in the dry season.

The majority of animals come to a waterhole during the afternoon and evening, especially when food is plentiful and the weather not too hot. But when food and water are scarce, feeding grounds for elephants may be more than 30 kilometres from water, so that the animals cannot return to drink until late at night or even until the middle of the next day. Each adult elephant consumes 140 to 170 litres of water, sucked up in the trunk 9 litres at a time, and squirted into the stomach with a noise like a cistern emptying.

It is obviously a great advantage to plains game to be able to make up water-loss quickly. The

donkey is able to make up a water-loss of twenty per cent of its body weight in less than two minutes. Other dry country ungulates are probably able to do likewise. The camel needs to visit water very infrequently, and can exist for a fortnight on only dry food. It can tolerate a much greater depletion in body weight than most other mammals, more even than the donkey, and may lose thirty per cent of its body weight without ill effect, compared with twelve per cent for most other animals. At the same time its appetite does not diminish with desiccation until this becomes severe. It is also able to make up the water-loss afterwards by an intake of water such as would kill most other animals.

The adaptation of the camel and donkey to a dry environment does not involve complete independence of drinking water, but rather an ability to economize in its use. Oryxes and Grant's gazelles are tolerant of very dry conditions, and are able to stay for long periods in waterless country during the dry season—an advantage that enables them to feed over large areas not available to the water-dependent species. Drought-tolerant ungulates like these obtain most of their water intake from their food (the dryest grasses found under natural conditions in East Africa contain three to five per cent water by weight) and lose very little water in their sparse and concentrated urine and droppings. Browsers get all the moisture they need from the leaves of bushes, which are hygroscopic and which, even during a drought, can consist of sixty per cent water. Antelopes feed mainly at night, when relative humidity is higher and leaves are wet, and they may obtain additional water by licking dew. In arid regions, where plant cover has not been lost, the volume of dew may be greater than rainfall. In one very dry area, 40 kilometres away from the nearest free water, impala drank dew in the early morning and went on drinking it in the shadow of trees until it had all evaporated. But in dry, over-grazed areas there is no appreciable dew because evaporation from the bare soil is too great.

A few animals are able to live in very arid bush without ever drinking standing water at all. Gerenuk, for example, have never yet been seen to drink in the wild. Gerbils also inhabit arid bush and semi-desert, and exist entirely without drinking. There are many gerbil species, ranging from very small mouse-size to large rat. All are nocturnal, sand-coloured, with big eyes. They are related to rats, but intermediate in appearance between true rats and jerboas, very specialized true desert-dwellers not occuring in East Africa.

Gerbils are much more mobile than rats, and probably lead a more nomadic existence. They subsist mainly on seeds, supplemented by bulbs, tubers and insects, all of which contain a fairly high proportion of water, Even the driest seeds, like the driest grasses, contain some water, and more is formed by the oxidation of food during metabolism. However, physiological economy of water is very necessary, so the animals must stay underground throughout the day. All species of gerbils are therefore inveterate excavators, often digging mazes of burrows about 15 centimetres below the surface.

Gerbils are the prey of nocturnal predators, especially those whose approach is stealthy, such as owls and small cats. Snakes may seek them in their burrows, and monitors try to dig them out, but a gerbil is often able to escape underground by digging further into the sand and closing the burrow behind it, or by bursting out of a stopped-up bolt-hole and running swiftly to another set of burrows.

Another small mammal of some very dry regions is the extraordinary naked mole rat. This lives entirely underground, only occasionally coming to the surface, perhaps to collect seeds or other vegetable foods, though normally it feeds on tubers, bulbs and roots underground. It is a baby rat-sized rodent; indeed it looks like a baby animal for it has no fur, only a few sparse bristles, a fringe of hairs around the edges of its feet to help it shovel fine soil, and whiskers on its lips.

Most burrowing mammals, for example, the golden moles and mole-shrews of the mountains, and the well-furred Kenya mole rat, live and work alone, but the naked mole rat is unique in that it lives in colonies of up to a hundred individuals which all co-operate in the digging of underground tunnels. Teams of naked mole rats form living chains, with three distinct working sections. The excavator does all the digging, biting at the earth with its big incisors, gathering it with the forepaws and kicking it to the animal behind. This animal collects the soil from the excavator, and then works its way back along the passage, beneath the rest of the chain gang. It leaves the load of earth with the final animal at the tunnel entrance and rejoins the excavator by crawling over the backs of the other members of the chain gang, who are moving backwards in their turn with loads of earth towards the entrance. The soil-disposer at the burrow

A small gerbil manoeuvres a stone backwards out of the entrance to the burrow it is digging. Gerbils dig rapidly into sandy soil, throwing the earth out with their hind feet. They spend the day underground to keep cool and conserve moisture.

An extraordinary rodent, and one of Africa's oddest mammals, is the naked mole rat. As its name implies, it has no fur, just a few tactile bristles; and no ear-flaps.

mouth vigorously kicks the earth out, straight up into the air. The soil is too fine and dry to form the usual type of mole-hill, so the spurts of dust fall to form typical naked mole rat volcanoes, that can be seen apparently puffing smoke when the excavators are working busily.

Several invertebrates are found in arid areas and exist without drinking. Under nearly every stone a scorpion may lurk. Dry country scorpions are usually rather small and finely-built, with brown bodies and yellow, almost translucent pincers; forest scorpions are often great greenish-black creatures with massive pincers. Desert

Scorpions are common in dry country. During the day they remain in the coolness and humidity of a burrow or beneath a stone; when they emerge at night they use their extended pincers as feelers to find their way and locate small animal prey.

The leg-like pedipalps of a solifugid are covered with long, fine, tactile hairs. Like the scorpion, this dry country animal feels its way about at night. Its jaws are the most fearsome of any invertebrate; the two pairs, side by side, work backwards and forwards to pulp up the prey.

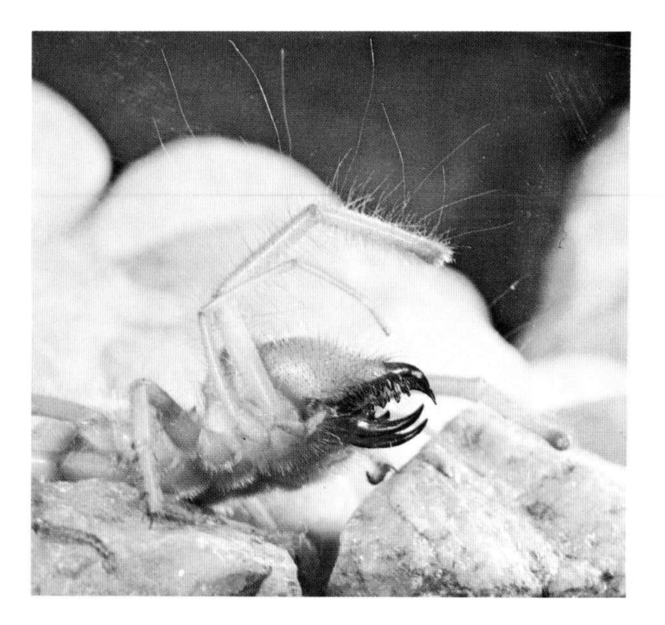

scorpions avoid desiccation by remaining below ground during the day, and can exist for very long periods without drinking or eating. But when they feed they can gorge themselves to double their fasting circumference.

Solifugids are another predatory invertebrate common in arid areas. Solifugid means "fleeing from the sun", and most species are strictly nocturnal. They, like scorpions, are wholly carnivorous, feeding on spiders, insects, smaller solifugids and scorpions, as well as mice, birds and lizards. They are rather horrific, reddish-golden, spider-like creatures, covered with long, stiff, tactile bristles, but they are not poisonous. They lack the scorpion's pincers for holding and its sting for paralyzing, but have two powerful pairs of jaws set side by side which work backwards and forwards to pulp their prey. Solifugids never need to drink; they obtain sufficient moisture from their food.

Among birds very few, if any, can exist entirely without drinking. Ostriches can go without water for many days, but they drink frequently when it is possible. A ground hornbill is known to have gone six months without drinking. But most birds, like most mammals, are not fully adapted to a dry

Like most other birds, the brimstone canary needs to visit water regularly to drink. It drinks by alternately sipping and raising its bill, tipping the water down its throat.

environment, and must visit water frequently to drink. In arid areas birds fly long distances to water. Flocks of pigeon-like sandgrouse flighting to and fro in early morning and evening are a familiar sight in semi-desert country. The birds may gather in huge flocks to drink, some having flown up to 64 kilometres to the waterhole. Here they gulp enough water to last them for the day. During the breeding season the cocks take water back to the chicks by means of a remarkable modification of the breast feathers which enables them to soak up and retain twice as much as normal feathers can carry when wetted. At the waterhole a cock sandgrouse walks belly-deep into the water, then crouches down and ruffles out his feathers, rocking to soak them. Even after flying twenty miles in hot weather he can deliver about half an ounce of water to his chicks, who nibble the feathers through their beaks to extract it.

Dust

Some birds bathe in water, not so much to cool themselves as to help maintain their feathers by the preening and oiling that goes on afterwards. Other birds never bathe in water. Instead they

The ground hornbill is one of the few birds known to be able to go without drinking for long periods. A bird of open country, it walks about on the ground, usually in family parties, snapping up dung beetles, millipedes, grasshoppers, lizards and fallen fruit.

A zebra enjoys a vigorous roll in a dusty depression. While the zebra is in such a vulnerable position a kongoni keeps watch for the possible approach of lions.

A mother warthog with four piglets and a companion at the salt-lick. Warthogs prefer to eat soil when it is dry; some animals, such as buffaloes, prefer it when it is wet.

bathe in dust. Guineafowl, francolin, bustards, sandgrouse and mousebirds flap and ruffle their feathers in it, even roll in it, kicking it over themselves with their feet. Afterwards they explode into the air in puffs of dust, like little rockets. Ostriches also dust-bathe, and raise a small dust-storm when they shake their feathers afterwards. Lammergeiers crouch in dust, and their breast feathers become stained according to the colour of the earth. Those stained russet with iron oxide were once thought to belong to a separate race from lammergeiers with paler breast feathers. These are all birds of open places where dust is more abundant than water. Others may dust-bathe at salt-licks or under overhanging rocks, anywhere where dry earth is accessible. Here hoopoes hollow out dusting holes in which they rotate till they are half buried; larks shuffle around with drooping wings and ruffled plumage; rollers, bee-eaters, hornbills and nightjars wallow in the dust.

Dust-bathing is not confined to birds; some mammals dust themselves. Gerbils, particularly, like to roll in fine loose sand, and groom themselves afterwards. Members of the horse family all roll. A herd of zebras will visit a dusty spot and repeatedly roll, kicking vigorously. Elephants also make great use of dust. After drinking they spray themselves all over with water, or they may wallow. They then very often take a dust-bath as well, spraying their bodies with fine earth, so that, like the lammergeier, they take on different colours according to the soil of the region. In areas of fine white alkaline soil they may be pale grey; in the black cotton areas dark; and bright red on laterite. These coatings of dust may have a camouflaging effect during the dry season, but in the rains bright ochre could hardly be more conspicuous against the green flush of new leaves.

A favourite place for animals to dust-bathe is at the salt-lick. This is often an extensive patch of bare soil where larger animals have mined the earth for the natural mineral deposits. Sodium chloride is not usually present in natural salt-licks; but the soil may be rich in trace elements which concentrate near the surface in arid regions, where there is insufficient rainfall to leach them. Herds of plains game visit the lick irregularly, not necessarily daily as they visit water.

Dust is not a habitat any animal might be expected to live in, yet ant-lions do so. Adults look like damsel flies with knobbed antennae; their larvae are among the few insects that make traps to catch their prey. The ant-lion's trap is a conical pit, in the bottom of which the larva lies hidden. Any insect walking on the edge of the pit starts a miniature landslide and is deposited in the waiting jaws. The larva can survive true desert conditions. Only the overhead sun at midday inactivates it in very hot weather, and it can live in dry dust without food for half a year.

During the dry season dust is an almost living feature of the landscape. It rises in clouds from beneath the feet of game and cattle, and guinea fowl kick up clouds as they scrape about to find food. At the same time there is the wind, not blowing steadily from any one direction, but in sudden strong gusts as air rushes from one local hot spot to another. Some very hairy moth caterpillars and beetle larvae roll themselves up and are bowled along the ground by the wind. A few plants, such as the small succulent *Carallumas* with parachute seeds, rely on distribution by wind; and some acacias are mainly wind-pollinated. Wind tosses the manes and tails of plains game, the crests and plumes of birds, and throws up waves and spume around lake shores. Large animals welcome the breeze for its cooling effect during the hot dry days, and for the dispersal of the many biting, sucking, or simply tickling flies that are otherwise a continual irritation to them.

As the ground heats up during the morning, columns of hot air rise. The wind sucks up the dust

A hen ostrich rising from her dust-bath. Many dry-country birds use dust rather than water to wallow in. The preening and oiling of the plumage that goes on after bathing, whether in water or in dust, helps to maintain the feathers in good order.

into spectacular dust-devils that roam the bush, drawing up soil, dead leaves, small insects and any small loose objects in their path. Dust columns may be 60 to 90 metres tall and roar for a kilometre or more across the plain before blowing themselves out. Small animals such as the harvester ants have to stop work in the windy season, as they are very easily blown off their roads by a gust. The mounds of husks accumulated outside their burrows during calm weather are quickly dispersed by the wind and the ants subsist on grass-seeds stored underground.

Much of the heavier material picked up by the winds is soon dropped again, but the finest particles are carried to high altitudes and transported great distances before they fall back on the soil. At the height of the dry season the whole land may be covered by a fine dust. Sometimes the amount in the air is so great that even at midday the sun is only visible as a pale disc through the haze, and an orange light suffuses everything. When smoke particles from numerous grass and forest fires are added to the atmosphere, there is almost an orange fog. At this time of year the sun goes down like a crimson ball, and the sunset colours may be amongst the finest anywhere in the world.

Fire: its effect on the habitat

Fire has been a major ecological force in Africa since time immemorial. It has been the traditional practice of pastoral peoples such as the Masai to set light to the plains during one or both the dry seasons. This has the effect of keeping down shrubby trees, such as the whistling thorn which is useless to cattle and would otherwise invade the plains; and it maintains the red oat grass and *Pennisetum*, making this sprout fresh growing shoots at a time when fodder for cattle is most scarce.

In some parts of Africa, particularly in hilly areas, annual bush fires have severely damaged and devalued the habitat. But in other parts, burning has produced stable areas covered with nutritious grasses and other perennials that protect the soil and provide excellent fodder. The vast plains and open woodlands which maintain huge herds of plains game—which in turn support the great carnivores and scavengers—have resulted from the prolonged effects of annual fires. Bush fires are today as important as topography and climate for the vegetation and fauna of a region.

The main effect of fire is to develop or encourage special fire-tolerant plants and animals which depend on periodic burning. Fire-tolerant trees have thick, corky bark; some produce seeds with a woody integument which explodes in heat, thus hastening germination. Many fire-tolerant plants have large underground bulbs or tubers and slender ephemeral parts above ground. These are known as *geophytes*; they often burst into flower soon after a fire when the ground is otherwise bare and black. Grasses sprout at once after a fire; they are able to survive burning because they have extensive systems of underground rhizomes from which they produce fresh shoots when the old growth is destroyed. New fire-induced shoots of star-grass and red oat grass are a better source of protein, calcium and phosphorus than the taller older growths, so they are preferred not only by domestic cattle but by the wild herbivores, which flock to a recently burned area and may seriously overgraze it. Other animals also flock in, but for differ-

Ant-lion larvae live in dry dust, in which they excavate conical pitfalls for ants and other small crawling insects. When it is ready to pupate, a larva spins a spherical cocoon around itself. Sand grains stick to the outside of the cocoon, effectively camouflaging it. After many dry months, rain penetrating the sand stimulates the ant-lion to cut through the cocoon. The adult insect emerges and hangs up to expand and harden its wings, leaving the old pupal skin, papoose-like, looking out of the cocoon.

Fire used by the pastoral peoples to clear bush and bring on green grass for cattle, has long been a major ecological force in Africa. Here kites converge on a fire to snap up small animals as they fly or run from the flames.

A pillar of dust whirls briefly across the dry plain, like a live thing. Constant trampling by wildebeest and other game coming to the waterhole has destroyed the grass cover and subjected the soil to wind erosion.

ent reasons. Marabous, kites, pied crows and others find small dead animals killed by the fire. Living prey is easier to catch with its cover gone, so smaller insectivorous birds also appear. Wheatears, black-winged plovers and Temminck's coursers are often attracted to burned ground.

Many plants have not developed a resistance to burning, especially forest plants which are not normally subjected to fire. Fire can make little headway in forest unless it is assisted by man or, as we shall see, elephants. Forest felling and clearance with shifting cultivation result in the spread of grass where none grew before. This allows fire to enter. Most vulnerable are forest edges and clearings. Once fire has penetrated, withdrawal of the rain forest can be rapid and is almost always irreversible, for fire prevents the regeneration of the forest trees. Forest is replaced by savannah, dotted with small fire-resistant trees. Energy is lost

The black-winged plover is attracted to burnt areas after a grass fire. Fresh, green grass blades are already sprouting from among the charred stubble of red-oat and Rhodes grass, and will be eagerly sought by the herbivores.

and the soil impoverished each time the vegetation goes up in smoke. With loss of ground cover, erosion is accelerated. With loss of water, climate is affected. In the end, whole regions may turn gradually from forests into deserts.

Vegetation is not the only thing that suffers from fire. With the destruction of their habitat, forest animals disappear and even in fire-tolerant habitats the annual destruction of small animal life must be colossal. Slow terrestrial creatures are the chief sufferers: tortoises and chameleons and giant snails which cannot move fast even to save their lives. In the dry season the bleached shells of giant snails can be seen in places scattered by the hundreds all over the bush, among the silvery skeletons of trees. The young of some ground-nesting birds are also at high risk. Ostrich eggshells are thick enough to withstand fire, and the young are soon large enough and strong enough to escape the flames by running; but the chicks of other ground-nesting birds are not. These species therefore usually nest at the time of lowest risk from fire, either at the beginning of the dry season when the grass is too green to burn, or at the end of the dry season when it is already burnt.

During a fire insects face an additional hazard to the flames, for flocks of opportunist birds are attracted by the easy pickings. Hundreds of kites and large swallows swoop through the smoke to pick off grasshoppers and other insects in the air, while tawny eagles, drongos, black-shouldered kites, marabous and others pounce round the fringe, snapping up rodents, snakes and lizards which are fleeing before the blaze.

It is possible that some rodents and other burrowing animals may escape being burnt if they remain underground. Generally grass fires are not very hot at soil level, although the temperature may sometimes be very high when there is a good layer of dry humus. So the grass nests of vlei rats in clumps of tussock grass are completely charred through, but grass rat burrows remain cool only a few centimetres underground. The rats that survive are probably able to find enough food—roots and fresh-sprouting grass-blades,—but they suffer from lack of cover in which to feed, and so are quickly finished off by predators. More mobile nocturnal species, such as the multi-mammate rat, are prepared to run the gauntlet of the predators and will cross quite large areas of bare ground to reach new living places.

The majority of conservationists are opposed to the use of fire, seeing it as the principal agent in the degradation of habitats over much of Africa. But soil and vegetation vary greatly and there are some regions where the effects of annual burning are beneficial, as in the Kidepo Valley of Uganda. Here the burning is carried out in a controlled way by biologists rather than by herdsmen, to prevent open grassland from reverting to dense bush, which is undesirable both for grazing herbivores and for the people who come to look at them. Not only is the area over which the fire spreads controlled but also the time of year at which the grass is burnt. Burning early in the dry season when the grass is still slightly green produces fires that are less fierce and therefore less destructive and more easily controlled.

Migration

Grasses and grazers, bushes and browsers, evolved at about the same geological time and have balanced each other ever since. When fodder is in short supply, the herbivores move to areas of better grazing or browsing. Migration is the natural device for making best use of land, and the majority of ungulates show some seasonal movement. During the dry season those species that are dependent on their daily drink, such as zebra, wildebeest, buffalo, eland, and Thomson's gazelle, concentrate within a day's walk of permanent water, while those that can do without

Two young wildebeest bulls spar with one another, clashing horns and raising a cloud of dust in what appears to be a fierce fight but is in fact a harmless trial of strength.

drinking, chiefly impala and Grant's gazelle, move out of that area and join oryx and gerenuk in the waterless places. Waterbuck may also move: they go not to dry places but to areas of permanent swamp.

The migration of wildebeests and zebras is the most spectacular of the seasonal big game movements in East Africa. From the short-grass plains of the Serengeti, half a million animals set out in May and early June and trek 240 kilometres to their dry-season grazing grounds where there is permanent water and plenty of shade trees. The herds travel in long columns, wending their way over the grasslands. While on the move, the wildebeest bulls maintain movable territories. They stay outside the herds of cows and young and defend their immediate surroundings against other bulls.

On the whole most antelopes are quiet creatures, but wildebeests are rarely silent. The bulls low, bleat, grunt and snort as they cavort about. The younger bulls engage in ritual pushing matches and trials of strength, but rapidly make off in a cloud of dust at the purposeful and lordly appearance of a dominant herd bull. Courtship in wildebeests, as in other antelopes, contains many stylized behaviour patterns. In one display, usually performed at some distance from the herd of cows, the bull rears onto his hind legs, presumably presenting a formidable appearance to his adversaries and incidentally a somewhat satanic appearance to a human onlooker.

During the dry season impala disperse through the dry, apparently inhospitable bush. The main reason for this is that they feed on the pods of certain acacias which ripen and fall during the dry season. The pods of most acacias are thin and papery, and burst while on the tree. The seeds shower onto the ground, the empty pods remain hanging on the tree. Browsing animals eat the pods, but do not pick up the scattered seeds from the ground below. However, the pods of *Acacia tortilis* do not burst. They are fleshy, large and heavy, and when ripe they fall to the ground without shedding their seeds. These pods have a strong smell which is very attractive to herbivores. Being rich in carbohydrates they form an important part of the dry season diet of impala and kudu, as well as of grass rats, which even put on fat at this season. In over-grazed areas where there is little or no grass left in the dry season cattle also eat the pods.

Impala, then, in some areas are dependent on the pods of this acacia for subsistence during the dry season. In turn, the acacia is dependent on impala for the dispersal and germination of its seeds. When ungulates chew the pods, most of the hard, smooth, rounded seeds are not crushed and are also unharmed by their passage through the animal's gut. Indeed, they do not germinate unless they have passed through the gut of a ruminant. Two more species are closely involved in this cycle: large mound-building termites and star-grass. Like impala, termites are dependent on this acacia for food; in some areas dead thorntrees are their staple diet. The large mounds of the termites are scattered all through *Acacia tortilis* country. On old or inactive termite hills grows star-grass, the grass which is most sought after by impala—who may manure it while eating it. Thus impala spread the acacia, the acacia feeds the impala and the termites; the termites encourage the star-grass, which also feeds the impala; the impala help to nourish the grass.

Elephants

Elephants join wildebeests and zebras near permanent water at this time. During the dry season they subsist mainly on the bark and twigs of trees and bushes near rivers or permanent waterholes. They use their tusks to gouge off the tough bark of thorntrees, and whittle twigs between their teeth to remove the bark. They chew wild sisal leaves to extract the moisture, and dig for tubers with tusks and feet. Of all large wild animals, elephants are the most adaptable in their choice of food plants, and feed on any of a hundred different varieties from head-high coarse grasses rejected by other grazers to delicacies such as underground bulbs. Much of what they eat is woody or fibrous, and because their digestive systems are not very efficient an elephant needs to consume about 150 kilos of plant material each day. With such an enormous intake the effect of large herds of elephants on their habitat is profound.

Sometimes the ecological influence of elephants may be beneficial. When they move into an area of swampland, by mowing and trampling they reduce a tangle of two-metre grasses to a freshly-sprouting greensward that other animals can utilize. Or they may sow whole new groves of *Borassus* palms by eating the fruit and depositing the unbreakably hard seeds several miles away.

Every part of the baobab tree is edible. During the dry season elephants attack the trunks, gouging out the fibrous wood. Few trees are left unscarred, and some have been ripped to pieces.

In other places, particularly in wooded country, the effects of elephants are definitely harmful, not only to the vegetation, but to game, including the elephants themselves. In the small isolated patch of high hardwood forest in the Murchison Falls Park, and in many *Terminalia* woodlands, the elephants damage and even kill the trees by browsing and ring-barking them. By trampling the undergrowth they let in light so that grass can now grow. Once grass is established, fires can penetrate the forest and complete the work of destruction. In drier thornbush areas, large concentrations of elephants, with the help of fire, rapidly convert the countryside from bush to grassland. Trees are pushed over and with the large accumulation of dead trunks and branches, and the invasion of the opened areas by grasses, fierce fires sweep through, destroying regeneration. Thus elephants, aided by fire, are the prime natural converters of forest to bush and bush to grassland.

Drought

The destructive influence of elephants is particularly felt in drought years when food is scarcest and competition for it keenest. During the very severe drought in Kenya in 1961 the large herds of elephants in Tsavo National Park were extremely short of food. Very little greenery was left along the Galana river. Herbs and shrubs were eaten down to the ground, and plants which elephants normally never touch were eaten. Mostly the elephants subsisted on the dry bark and leafless twigs of trees and shrubs which, in spite of their unappetizing appearance, maintain a relatively high nutritional value. Destruction was far greater than in a normal dry season, and very heavy browsing and ring-barking killed hundreds of thousands of trees. The elephants also destroyed baobab trees at an unprecedented rate. They have not always eaten baobabs, but pressure of population here and elsewhere has caused them to turn to this new food. In the late 1950s they started attacking baobabs, and now few large trees remain unscarred, while many are in process of being spectacularly destroyed. All parts of a baobab are edible, so that not only are branches, bark, leaves,

An elephant with her young calf leaves a small seasonal waterhole. Trampling and puddling the wet mud as she drinks, she helps enlarge and maintain the pool, which, as a result, will hold water for longer into the next dry season.

flowers and seeds eaten, but whole large trees, which may be a thousand years old, are ripped apart for the moisture and calcium content of the fibrous wood. Fire completes the destruction by killing the seedlings. Almost as if in retribution, there were several instances of elephants being killed by the falling trunks of the baobabs they were eating.

It is not known how browsing antelopes such as kudu and gerenuk suffered in competition from the elephants during the 1961 drought. Being nomadic to some extent, they were probably able to move out to less heavily browsed areas which the elephants could not use because of lack of water. But the black rhinoceroses suffered very severely, to the point of starvation. Along a 64 kilometre stretch of the Athi river alone, 282 rhinoceroses starved to death, though it is not certain that the elephants were wholly to blame. Black rhinoceroses never move from their home range in the dry season, even when competition from elephants is severe. If their territory does not contain permanent water, they do without rather than make long treks to it as elephants do. Rhinoceroses are particularly selective over their food in the dry season, and will eat only the green parts of plants in preference to dry withered bits, to obtain the moisture they need. They also chew succulent latex-bearing plants, wild sisal and euphorbia. Stems or leaves may be chewed for as long as twenty minutes to extract the moisture, after which the fibres are spat out. Finger euphorbia is eaten in great quantities where it is common, and much sought after where it is less common. To get at the higher branches, a rhinoceros stands up on its hind legs and climbs with its front feet as high as it can. It then hooks its horn among the branches, and walking backwards, snaps off large amounts of tree. It eats all the small branches and gouges the bark off the bigger ones with its front horn and teeth.

Finger euphorbia is not indigenous to East Africa, but was introduced from India. Its colonization and spread throughout the region must have had a profound influence on the survival of the rhinoceros in arid areas without water. Another introduced succulent, the prickly pear cactus, of which four species have been brought from South America, may also have influenced the survival of some animals in very dry areas. The thornless type is a valuable dry-season fodder for cattle, but the thorny type is more widespread as it has been much used as a hedging plant. In a few years, it develops into dense thickets, which provide cover

During the dry season the black rhinoceros selects green leaves and twigs wherever it can, but during severe drought it must subsist on dry twigs and bark and such moisture-rich though bitter succulents as are available on its home range.

Grazing is almost gone in this drought stricken area, where rain has not fallen for almost a year; but this herd of eland does not look distressed. They can subsist quite well on the protein-rich twigs of acacias. The prickly pear cactus is an introduced weed, but in districts like this may provide fodder for animals otherwise desperately short of food.

The Kei apple is another introduced hedging plant benefiting many animals during lean periods when indigenous food is scarce. Long-tailed thicket rats live on the apples and nest in the dense thickets which this small tree provides.

for many kinds of small animals. Its spines deter creatures from eating it except in time of direst need, when it is chewed by the larger ungulates. However, it is of greatest value to smaller creatures and may provide their only means of subsistence in a prolonged dry season. During a severe drought grass rats and small birds were seen to be eating not the cactus itself, but the cochineal insect that was introduced in attempts to control the cactus: pink droppings all around the cacti showed what they had been feeding on.

Other introduced hedge plants may also provide locally abundant food during a lean period. The Kei apple produces a bountiful crop of delicious-looking small, yellow apples which ripen and fall in the dry season. During the day the fruits attract flocks of glossy starlings, mousebirds, coqui francolins, and brimstone canaries, whose yellow breasts match the yellow apples; while at night they are eaten by small game such as steinbuck and thicket rats.

In a severe drought animals are often forced to make do with unpalatable or unusual alternatives to their diet. Warthogs dig for roots and tubers with

their tushes when grass is scarce; rock hyraxes chew the stems of the large *Caralluma* which even goats do not touch; bushbabies and pottos eat tree gum. Even lions may chew the stems of wild sisal to obtain moisture, although in a normal dry season they get enough water from the stomach contents of their prey. Cheetahs, too, lap the blood that collects in the rib-cage of their kill as in a bowl, and obtain sufficient moisture that way.

In a normal dry season the wild animals of Africa do not suffer hunger or thirst. Over millenia they have adapted to heat and periodic drought. The animals that really suffer are the cattle, especially during drought years. Cattle need to drink at least once a day, which means there is a limit to the distance they can graze from water. When many waterholes dry up, all the cattle herd around the few that remain, grazing out all the grass for miles around. Other grazing animals suffer from the concentration of cattle. Worst hit are the zebras, which under normal conditions come through a drought remarkably well, since they can digest coarse dried grass which is useless to other animals. During the 1971 drought, however, zebras in the Maralal Game Reserve were dying in hundreds. Yearlings suffered most; older beasts and foals at foot fared better. The Samburu cattle were also dying, for there simply was no grazing. Only the goats looked sleek and fat, and it was a sad sight to see hordes of them fanned out over the hillsides, nibbling down to the bare earth what little stubble remained. They also ate the protein-rich twigs of thornbushes denied to cattle and zebras, but fortunately available above goat

A zebra carcass, only half cleaned by vultures, is left to dry in the sun and wind. Perhaps the grass will grow more richly around this corpse when the rains finally come.

Yearling zebras fare worst in a drought. Jackals have eaten the hindquarters of this carcass, leaving the rest as a feast for Egyptian vultures.

Left: With their grazing destroyed by the domestic herds of the Samburu, these Burchell's zebras are starving. They do not, however, appear skeletal like the cattle: their contour-striping and quite thick, rough coats obscure their thinness.

level for the taller wild herbivores, giraffe, eland, impala and gazelles.

The drought of 1961 is said to have killed three-fifths of the cattle in Kenya, and the drought of 1971 also caused severe losses. The problem of survival in a drought is made worse when, as often happens, it is preceded by several years of good rains. Then the grass is plentiful and the cattle increase. To supply the big herds with water, new wells are dug. But every year the grazing shrinks. A larger number of big animals can quench their thirst, but they cannot find food, and the pressure on the vegetation destroys it further. In places, more than ten times the number of cattle are maintained than the country can support in times of drought. In some areas the proliferation of artificial wells has diverted water from what used to be permanent waterholes for wild animals. The result has been a reduction in game because, over wide areas, the amount and distribution of permanent water is the limiting factor determining the number and distribution of game.

During a severe drought, there is a feeling of suspended animation in the parched plains and the grey bush, an air of simply existing if possible, rather than living. As pools dry up and turn to pans of cracked mud, crocodiles lie lethargically waiting for the water to return. Insects and other invertebrate life seem to have died out, yet flocks of starlings and troops of baboons are finding something somewhere to sustain them—perhaps a few termites under still-damp elephant dung or a solitary wolf spider waiting in its burrow. Only the carnivores and scavengers have plenty to eat as carcases of dead zebras and cattle litter the plains. Many are left to dry in the sun: they are too numerous for the scavengers to dispose of. Throughout each long, hard, hot, dry day there is a feeling that all animal life is simply enduring and waiting; waiting for clouds to begin gathering, for humidity and softness to return to the air and the earth, for the rains that will make the scorched and barren countryside spring to life once again.

The Wet Season

The coming of the rains

On the equator, rain falls in two seasons, just after the equinoxes in April and November. Throughout East Africa these wet seasons are modified by distance from the equator and from the sea, and by altitude. No part receives rain evenly distributed throughout the year; one of the two wet seasons is longer and wetter than the other. In many places the total rainfall is less than 50 centimetres a year, grading to as little as 25 centimetres in the northern deserts. In these places the rainy seasons are separated by very long droughts. When the rain falls, there is sufficient only for a brief flush of annual herbs and grasses, and a short intensive breeding season for birds and other animals. In other places there may be 150 to 175 centimetres a year and in the mountains much more, with hardly a month in which there is no rain. The vegetation of the higher rainfall areas used to be forest, but today much of it is wooded savannah, even grassland. Here the differences between the seasons are not dramatic. Flowers bloom the year round, vegetation remains green and birds nest at any season. But in the lower rainfall areas the change is very dramatic, from parched inactivity in the dry season to intense activity, flowering and verdure in the rains.

Rain falls in the tropics, as elsewhere, when winds converge. Air rises above an area of convergence and cools by expansion so that moisture in it condenses, then precipitates as rain. Two main convergence zones affect East Africa. The largest and most important is the equatorial trough or *intertropical convergence zone* (ITCZ), a belt where trade winds or monsoons from the southern and northern hemispheres meet. This zone brings rain as it moves seasonally northwards and southwards following the sun, but is much modified by smaller, more complex, local wind patterns. The ITCZ affects much of East Africa, but rain is also brought to the Western Rift by a second semi-permanent convergence zone, the *Congo air boundary*, where Atlantic westerlies meet easterlies from the Indian Ocean.

Towards the end of the long dry season clouds begin to build up daily as the atmosphere becomes perceptibly humid. In response, every acacia bush bursts into blossom and leaf, filling the oppressive atmosphere with refreshing scent and providing welcome fresh browse for herbivores. The acacias' early flowering allows wind or insects to pollinate them before the first heavy showers batter the flowers to pieces. Now bright colours return to the bush. The flush of new leaves in acacia bushland is emerald, but other trees flush red. The *Brachystegia* woodlands of southern Tanzania break first into spectacular bright red or copper leaf, then gradually become duller, through olive to their normal glossy bright green.

When the long-awaited rain comes, it does not fall as a gentle drizzle but typically as a sudden heavy downpour, often in the late afternoon or just after dark. Everywhere there is great excitement at the coming of the rains. Even when the fall is many miles away zebra families congregate into large herds and the stallions run about braying excitedly before they all set off towards the storm. They have been seen galloping with wildebeests towards a storm 8 kilometres away, and will

Even before any rain has fallen the increased humidity of the coming rains stimulates acacias to blossom, providing copious nectar for insects such as the blue pansy butterfly fluttering here among the flowers.

Above, right: The first storm of the season, coming across the lake in the late afternoon, whips the murky water into wavelets. The flamingos become very excited, going through the motions of bathing in the downpour and taking off across the lake, apparently beneath the very arch of the rainbow.

Right: The mean surface winds during January and July in Eastern Africa.

Previous page: A young gerenuk buck among the spring-like greenery of fresh leaves and grasses at the beginning of the rains. The first showers of the wet season fell only days before, transforming the dusty, grey bush almost overnight into an emerald paradise.

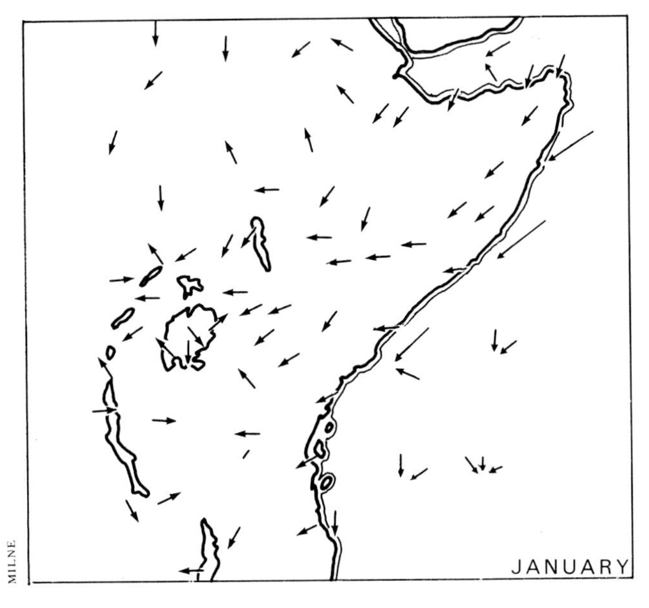
JANUARY
MILNE

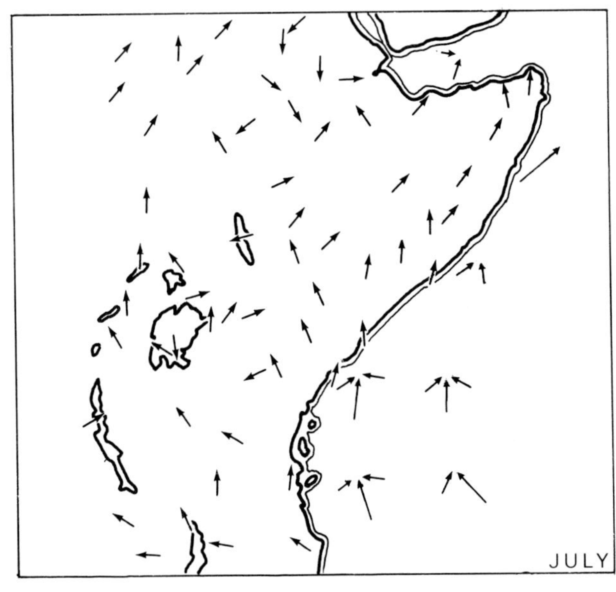
JULY

travel 40 kilometres in a night to reach a more distant one. Grant's and Thomson's gazelles will cover 15 kilometres in a night to reach the site of a first rainfall. During the rain great grown lions may splash and gambol in it like cubs. Wild dogs become very excited and run round licking the rain from each other's coats. Birds welcome it as a chance of a showerbath. Hornbills and sunbirds particularly, which do not normally bathe in standing water, flutter with great excitement in the downpour and among the rain-soaked vegetation. Parrots are specialist rain-bathers, and have characteristic spread-eagled and fluffed out postures for rain-bathing. Pigeons and doves rain-bathe in the same posture as they sun-bathe, lying over on their side with one wing extended and raised to let the rain get to their flank and underwing. Larks lie out in the rain with their wings spread. Even normally water-bathing birds become very excited during the showers and make clumsy attempts to rain-bathe.

As soon as the first rains have fallen in the bush many plants burst spectacularly into blossom. Plants with large underground bulbs or tubers flower within days of the first rain: delicate white lilies push up through the hard bare earth, and the

Birds become very excited during the first storms. A female bronzy sunbird fluffs out her feathers and flutters her wings as she enjoys the shower bath.

Rainwater floods down into the soda lakes from the surrounding hills. Lake birds such as black-winged stilts and a ringed plover gather at the river mouths to bathe excitedly in the pools of water.

After the main downpour this speckled pigeon came out from shelter and just stood out in the rain, allowing the drops to settle on its plumage. Afterwards it preened the more vigorously for having got its feathers damp.

In arid areas where there is little grass left at the end of the dry season, tiny fragile-seeming clusters of white lilies spring up out of the hard ground as soon as the first rains have fallen. Small beetles, such as this longicorn, and tiny chafer are attracted to the flowers and soon spoil them by eating the petals.

The desert rose looks like a small tree; it grows in arid regions and stores moisture in its bulbous lower trunks. Its pink blossoms are particularly lovely against the new greenery of sprouting acacia.

Carallumas, found only in semi-desert, are the African equivalent of small cacti. At the beginning of the rains their stems swell again with moisture, and buds appear. The large five-petalled waxy blossoms burst open suddenly with a distinct little "pop". Like much other succulent greenery at this time of year, stems and flowers are soon smothered in rapidly-reproducing aphids. The tiny gerbil is holding its hind foot while it washes its toes.

The beautiful, large-flowered Crinum lily appears in grasslands soon after the first rains, its flowers opening in succession and quickly dropping. The mushroom-like agaric may indicate a termites' underground fungus garden.

lovely deep pink blossoms of the desert rose open. Aloes send up tall spikes of orange trumpets, providing an abundance of nectar for sunbirds and bees. Small succulent carallumas that have escaped the goats in thornscrub thickets put out surprisingly large, starlike waxy blossoms, and giant euphorbias are covered with tiny, sulphur-yellow flowers.

The most surprising plant growths in the bush are the fungi, which force their way up out of the earth a few days after rain. Often toadstools appear in "fairy rings" around termite towers. They are the *agarics* or fruiting bodies from fungus gardens cultivated underground by the termites. Worker termites make elaborate sponge-like combs out of undigested cellulose on which their special fungus grows. They do not eat the fungus itself, except for the small white fruiting bodies or *conidia* which may be a valuable source of vitamins; but by providing warmth and moisture the gardens play an important part in air-conditioning the underground nests. The toadstools are a second form of fruiting body, but only appear seasonally. They may spring up anywhere in the bush as well as around termitaria, and locate otherwise unlandmarked termites' nests beneath the soil.

On the plains, too, many plants come into flower and the grass begins to sprout, transforming the blackened or straw-coloured land. Elephants reverse their dry-season movements and wander away from permanent water, spreading out through the bush where they rely on numerous small wet-season waterholes. They now eat vast quantities of fresh green grass, not indiscriminately, but selecting certain species and ignoring others, plucking the upper shoots and discarding the roots and stalks. They become very excited and push over trees, particularly *Commiphora*, without feeding on them. This strange behaviour may simply be playful trials of strength when they are in a general state of excitement.

Baby mammals

With the abundance of good grazing, most other ungulates move out from the permanent-water areas where they had concentrated during the dry season and disperse widely over the countryside. At the same time impala, waterbuck and Grant's gazelles, which had dispersed in the dry season, move back in large numbers. Many give birth now. Smaller antelopes have two breeding cycles a year, but most of the larger plains game have one annual breeding cycle and drop their young at the

Kongoni or Coke's hartebeest are most likely to produce their young while the grazing is best, early in the rains.

start of the long rains, when feeding conditions are at their best.

With the coming of the rains the great herds of wildebeest migrate back to the plains to calve. The way is led by Thomson's gazelles and zebras, which may be more sensitive to barometric pressure or to wind and temperature changes, and so are aware when distant rain has fallen. The long columns of wildebeest follow. The wildebeest return to their regular calving ground, for instance in the Ngorongoro Crater and the Loita plains, when these are sprouting with fresh green grasses. The herds follow well-defined traditional tracks making deep gullies across ridges. On the calving grounds all the cows give birth within a short period of a few weeks. Parturition takes place in the open and is usually very quick. The calf may be on its feet within minutes of birth, and will at once try to suckle, although perhaps from the wrong corner of its dam. Within only five minutes of its first wobbly efforts at standing it can run with the nursery herd. Once on its feet and moving it is able to keep going and keep up with its mother. Jackals may wean their cubs on the wildebeests' afterbirths, so plentiful are they at this time.

Other plains game, such as hartebeest and

Elephants become particularly excited during the first storms of the rainy season, and go around pushing over small trees—apparently just for fun.

Beisa oryx on the sub-desert steppes of northern Kenya. Only days before, these plains were parched and brown; the first rains coincided with this calf's birth and produced a flush of good grazing for the lactating mother.

A baby steinbuck is left hidden down an old ant-bear burrow or in the grass, with which it blends perfectly. Steinbuck and other small antelope fawns may be born at any time of the year, since they have two annual breeding cycles. Even so, a high percentage of them are likely to be born now when conditions are best.

zebra, may also calve when grazing is best, so that they coincide with the great wildebeest birth peak. Most antelopes are born into nursery herds, but zebras are born into family groups of a stallion, his mares and their last year's young. The stallion will actively try to defend his family from hyenas and wild dogs. Many antelope and gazelle mothers are also exceedingly brave in defence of their young. Oryx are a match even for lion, and a Thomson's gazelle will charge the smaller predators—jackals, baboons, martial eagles or secretary birds—sometimes succeeding in driving them away. At birth a gazelle fawn's coat is a much darker brown than that of the mother, and it will try to escape detection by lying motionless close to the ground.

The camouflage of baby antelopes is probably coincidental. Many lack the disruptive white spots of deer fawns, but their colours blend effectively with their surroundings. The brown baby coat of the gazelle fawn, for instance, blends well with old or new grass. But even so, during the breeding season a very heavy toll of fawns is taken by leopards, cheetahs, wild dogs, hyenas and jackals; and occasionally also by lions, servals, caracals and the larger birds of prey.

In the same spring-like season, vervet and patas

A baboon baby at first clings upside down under his mother's belly, but by the time he is four months old he can ride jockey on her with confidence.

monkeys and baboon babies are also being born. At birth, the ears of baboon babies are oversize, crinkly and pink. Their faces are pink, their muzzles corrugated and foreheads puckered. Pink faces and very dark body fur arouse strong emotional reactions in adult baboons, so that babies become the centre of interest for the whole troop, mothered by all the females and fiercely defended by the adult males. Like antelopes, monkeys normally give birth to only one young at a time. The need of antelopes to produce young fully active at birth precludes the development of more than one at a time, or very occasionally twins. With monkeys, it is an adaptation to life in the trees. The mother must be able to move about freely, even while carrying a new-born baby, and the baby must be born able to cling tightly to her, to leave her hands free for climbing and feeding. A young monkey's age can be told by the position in which it clings to its mother. At first it clings beneath the belly, with arms around her and fingers and toes tightly locked in her fur. It peeps out, upside down, between her front legs. A few weeks later it will try to ride on her back, but cannot yet sit upright. Not until it is about four months old can it ride jockey with confidence.

Baby vervet monkeys are also most likely to be born at the beginning of the rains. At first the baby clings to his mother's belly, and peeps out upside down between her front legs when she is on the move. When she sits down to feed or rest he is already in the most convenient position for suckling.

Zebra mice babies in their grass nest. They are born completely helpless, their eyes and ears closed, with a faint covering of downy fur. They will need to be kept warm by the mother for two weeks before they are ready for short excursions.

Like monkeys, bats carry their one baby clinging to their fur. But single births are unusual among mammals as a whole; multiple births are the general rule. The record for litter size among African mammals is held by the multi-mammate rat, which may give birth to any number up to twenty at a time. The female has a very large number of teats, up to twelve pairs, whereas other rats and mice normally have no more than five pairs. The babies are born blind and naked, but develop rapidly until by the time they are twenty days old they can fend for themselves. The mother can give birth every twenty-five days, while her babies can reproduce at two months old, so the multi-mammate population can expand five-fold every month during a good rainy season. In exceptionally wet years, when the breeding season is prolonged, the population may build up to plague proportions. Grass rats may also build up their numbers at the same time, and can be seen out in the daytime feeding on the new grass and scampering across open places, while multi-mammate rats come out at night to feed, and may be heard rustling and sqeaking everywhere.

Litter size in mammals is usually inversely related to the stage of development of the young

Spiny mice are unique among rats and mice in giving birth to precocious young. This baby, only one day old, is well furred though not yet prickly; and its ears and eyes are open. It has staggered from the nursery, but its mother has picked it up and is taking it back to the nest.

The northern Uaso Nyiro River, northern Kenya, in spate after rain, carries along with it tons of surface soil that has been eroded from the surrounding land during the violent downpours of the last few days. Only a week ago, the river bed was almost completely dry and the crocodiles were starving. In another day, as its level falls a little, the crocodiles will find plenty of corpses of animals left by the flood.

Acacia seeds germinating after rain. Their green parts will grow slowly, but their roots push down fast to find underground water before the soil dries out again.

at birth. At the other end of the scale from the multi-mammate rat comes the spiny mouse, which is unique among rats and mice in giving birth to fully active, precocious young. A female spiny mouse has a very long gestation period for her size, about thirty-eight days. The two or three precocious babies in a litter have their eyes open at birth, and can totter around the nest area straight away. They are well furred, but not yet prickly. In a few days they are nimble and active, at a fortnight weaned, and sexually mature, ready to reproduce, when they are about seven weeks.

Flash floods

As the wet season progresses, rainstorms become more frequent and are often very violent, with lightning and thunder. Animals are no longer excited by the rain. Birds of open country caught by a heavy downpour adopt a special rain posture, standing upright with the feathers sleeked against the body so that the rain runs off. Large animals shelter under trees, standing close together. Plains game stand with their rumps towards the worst of the storm, their backs glistening characteristically. At high altitudes, even on the equator, these very violent rains sometimes fall as hail, which shreds vegetation and does much damage in agricultural zones. When the storm has passed, it may leave whole areas of grassland flooded, so that for a time the problem is one of too much water where previously there had been too little. Herons, ibises and storks wade about feeding on the small animals flooded from their burrows.

In grasslands, the permanent plant cover provides good protection for the soil during violent storms and floods. Forests, with their tall trees and undergrowth and complex root systems are even better protectors of the soil. But where land has been denuded of vegetation by over-grazing or cultivation, violent rains are a powerful agent of soil destruction. Only part of the downpour penetrates the soil, the rest runs off downhill, forming small torrents which carry away humus and soil. In hilly country, the run-off scores deep parallel gullies; at every rainfall the gullies deepen until in a few years they may have eroded into seventeen-metre ravines down which the flash floods roar, carrying with them surface soil that may have taken natural vegetation thousands of years to lay down. On gentler slopes, sheet erosion is less spectacular, since at first it shows only as a change in the colour of the soil. Gradually, as the finer particles are removed, a layer of pebbles is left at the surface. Bare soil lacks the power to absorb and hold water, and evaporation from it increases. In this poor leached soil, baked hard and sterile by the sun, only a few desert-adapted plants can grow.

One of the adaptations to an arid environment which enables acacia trees to flourish in semi-desert is the development of very long root systems which can penetrate deep into the soil to tap moisture unavailable to shallower-rooted plants. After a heavy shower acacia seeds that have been lying on the bare earth beneath the parent tree get swept up and carried along by the run-off. The soaking causes them to swell and germinate. The aerial parts of the seedlings develop slowly, but the roots grow much faster. However, the moisture does not penetrate far into the ground, and it quickly evaporates, so that the seedlings will find their roots trying to penetrate waterless earth and most will die before the next rainy season. It is not clear how the seedlings that survive are able to do so, or how they can grow through moistureless soil to attain the root-lengths of the adult plants. The problem regarding drought-resistant trees is not how the grown trees survive, but how the seedlings ever reach maturity.

Much of the rain that falls on the thirsty bush country scarcely benefits that countryside at all, for only a small proportion of it soaks into the ground. The rest runs off in rills and gullies into temporary water courses. These disgorge into the big rivers which become raging torrents, bearing along trees, dead animals and silt. Under the

Lake Nakuru, strangely deserted of its flamingos and fishing birds after a rainstorm had caused a spectacular fish die-off there. Little egrets were one of the few residents that remained, while kites flocked in by the score to feast on the dead tilapia.

The flotsam of dead fish along the shore. Marabous gorged themselves on the corpses before they became too rotten in the hot sun for even these undiscriminating scavengers.

pressure banks collapse, especially where in the dry season elephants have dug beneath them in the dry river bed to find water in the sand. Several of the big rivers, for example the two Uaso Nyiros in Kenya, lose themselves in vast swamps, while others drain into lakes which may themselves have no visible outlet. Other rivers have their outlet in the sea. So by evaporation or run-off most water is lost to the soil.

When a very heavy downpour occurs in the region of Lake Magadi or Nakuru a spectacular die-off of tilapia may result. Probably an algal flush follows the influx of fresh water, and this leads to deoxygenation of the lake water and suffocation of the fish. At Lake Nakuru half a million dead fish were recently washed up around the shores after a rainstorm. Simultaneously all the fishing birds departed from the lake, and the flamingo hordes left overnight. The lake then had a most strange appearance, almost deserted of birds except for dabchicks and a few egrets, and an exceptional number of marabous, pied crows, kites and other avian scavengers which descended on the shore to feed on the sun-baked fish. But half a million tilapia are, after all, only a week's food supply for all the fish-eating Nakuru birds. The surviving fish soon built up their numbers and pelicans, flamingos and all the other birds returned to the lake in as great multitudes as before.

Giant red velvet mites can survive completely arid conditions for long periods. At the start of the rains myriads of them reappear, creeping over the ground in desert places for a brief, intensive feeding and breeding period.

Animals of the rainy season

During the rains the sky is overcast, with low oppressive banks of grey cumulus. Showers may fall all around, even if they do not hit the same area again for some weeks. The atmosphere remains humid, and the sun's evaporating power tempered by cloud. Rain means the start of a new life for a large variety of animals that need a damp atmosphere in which to carry on normal activity. At the beginning of the previous dry season these animals burrowed into the ground before it became too hard, and remained there throughout the hot dry weather (page 97). Now the rain softens the ground, awakes them from aestivation, and they dig their way out for a short period of intensive feeding and breeding.

One of the most conspicuous invertebrates to appear after rain is the centimetre long velvet mite or "red spider". They creep over the ground, often in myriads in really arid country, their scarlet colour a warning that they are distasteful. Birds, big spiders, even the voracious scorpions and solifugids leave them alone. While the ground is still damp they are diurnal, but as the atmosphere grows less humid they become crepuscular. They mate and lay their eggs within a few days. The young grow very rapidly, feeding on vegetable detritus, then go into aestivation, perhaps for a whole year or more, until the next rains.

Other normally nocturnal creatures that suddenly appear everywhere during the day are very large woodlice. These abound in rather shaded places such as among the leaf litter in riverine or gallery forest. They creep about in broad daylight for a few days, looking for mates, then revert to their normal nocturnal habits. Woodlice are crustaceans, an order whose other members live underwater or in very damp places. Lacking an efficient waterproof skin they are particularly susceptible to desiccation, so only come above ground by day after very heavy rain.

Multitudes of millipedes appear in the bushlands at this time. There are many sizes. The giants are splendid cylindrical creatures about 120 millimetres long, exquisitely precision-built, moving smoothly forward over the ground with synchronized waves of their 228 legs. All millipedes trundle about seemingly unaware of anything beyond their own antennae. Their assured be-

Millipedes are normally nocturnal, but at the beginning of the rains they may appear by the thousand above ground in the daytime, looking for mates. After pairing they disperse underground again and revert to their nocturnal habits.

haviour suggests that they too, are distasteful, and when alarmed they spring into a coil and exude an astringent substance. But millipedes are a favourite food of the banded mongoose, whose occupation of a termite mound can be diagnosed by the presence of quantities of millipede rings in the mongooses' latrines. Scorpions too, eat millipedes and in spring-cleaning their burrows bulldoze out mounds of millipede rings and earth which they tamp down outside. When a millipede walks by the entrance to a scorpion lair it causes a landslide which deposits it in the scorpion's underground dining-room.

Millipedes may sometimes occur in very great abundance. One species, a glossy black animal with orange legs, sometimes swarms twenty or thirty to the square metre. Some individuals may be feeding, scavenging on decaying plant material, but most of them are proceeding purposefully about looking for a mate. They are so thick on the ground one cannot avoid treading on some; elephants leave a trail of crushed bodies, and the sharp hooves of antelopes cut many in half: they still continue walking without their tail ends. A few birds eat millipedes—fiscal shrikes, red-billed hornbills and cuckoos particularly. After three days

Giant snails can remain dormant for well over a year, cemented into their shells by a cap of dried mucus. When rain softens the cap they emerge, to munch green leaves.

the millipedes disappear, not eaten but gone to ground, under logs and stones and elephant dung.

Rain puddle fauna

During the rainy season the downpours fill up all those seasonal waterholes that have remained dust bowls through the drought. Water is now available over a wide area of bush for all the big game, though even those ungulates that normally drink daily require little water during the peak of the rains because of the high moisture content of the lush greenery they eat. When they do visit the waterholes their trampling maintains and enlarges the pools, so that they hold the water for longer into the next dry season. The effect of animal maintained pools on the ecology is very great, since they enable animals to stay in the area well into the dry season. These pools also provide food and drink for butterflies, which suck up moisture and mineral salts from the wet mud where big animals have urinated; and they provide breeding grounds for a host of smaller creatures, insects and other invertebrates, amphibians and fish.

Some seasonal puddles, dry for half the year, miraculously swarm with little fishes during the rains. These are killifishes, amongst the most brightly coloured of all fresh-water fish. They grow to a length of only 40–50 millimetres and the males in breeding colours are living jewels. One species is olive green spotted with pale blue, red and white, with an all-red tail fin. The female is a drab grey-brown. Another is a lovely deep violet, spotted with brilliant red, with yellow pectoral and ventral fins. They live in rainwater puddles and seasonal pools and are excellent jumpers, flipping across the mud to find deeper water when their own pool dries out. When all the pools are dry the killifishes die. Their species is perpetuated by drought-resistant eggs.

Killifish eggs pass the dry season in the dried mud of what was once a pool. When the weather breaks and the puddle fills, most of the eggs begin to develop and hatch within hours. A few eggs do not develop with the first wetting, a safeguard should the rains fail and the pond dry up again too soon. The fry grow extremely fast, feeding at first on the

The male of the tiny East African killifish *Nothobranchius guentheri* is among the most colourful of freshwater fishes; the female is more drab. They live in seasonal rainwater pools. When the pool dries out individuals of a season die, but their eggs survive the drought and hatch in the following rains.

tiny crustacea which also hatch from drought-resistant eggs. As the fishes grow they graduate to larger crustacea and mosquito larvae. In only six to eight weeks they are sexually mature and start breeding. The males are extremely pugnacious, fighting one another and chasing the females. When a ripe female is ready to spawn she allows the male to swim alongside and wrap his dorsal fin over hers. Together they plunge into the mud, disappearing in a cloud of silt. They continue to spawn throughout the short breeding period until the pool dries up and they die. The eggs cannot hatch in the same water in which they were laid. They must go through a dry period of many weeks—twelve is the optimum, but they will readily survive much longer—so there is no danger of the fry hatching out and dying prematurely with their parents.

Temporary puddles have an abundant plankton of small crustaceans, known as "water fleas", which often occur in such millions that the water is soupy with them. Slightly larger crustaceans, like little oval bivalve shellfish, up to 3 millimetres long, are also sometimes abundant in a pool. These are clam-shrimps. They look like tiny translucent molluscs, for their shells completely enclose the body and limbs; but antennae and legs can be extended between the valves to beat and kick the animal through the water. They swim in little jerks like water fleas, scavenging anything microscopic, plant or animal, in the water.

A much larger kind of crustacean that flourishes in puddles is the fairy shrimp, a very primitive animal. Like the lungfish (page 97), it is a living fossil, being similar to very ancient fossil crustacea. Adults are about 2 centimetres long and swim on their backs like the related brine shrimps. They, too, feed by filtering microscopic particles from the water. They are quite beautifully coloured, mostly transparent but with some red on them, and with iridescent greens and blues on the body. The eggs are carried by the females in large egg-sacs. Eggs that have not been dried out may hatch, so that several crops of fairy shrimps can be found in the same puddle. If the puddle dries, however, the eggs can remain viable in the dust for years, and dried eggs actually hatch quicker than those that have not been dried.

Fairy shrimps and water fleas are very important in the ecology of the rain puddle and form a major item in the diet of the next class of puddle fauna, the carnivores. These aestivate through the drought and appear suddenly as soon as rain has fallen. One of the most conspicuous is a large water scorpion, a

big, wicked-looking brown bug, with a long slender respiratory snorkel-tube at the end of the abdomen, and mantis-like front legs. It normally flies by night but at the start of the rains it appears all over the bush by day as well, scurrying along in search of a puddle. Water beetles also appear; they are clumsy on land since their legs are oar-like paddles adapted for swimming. Oval in shape and beautifully streamlined, with smoothly polished dark green elytra, these beetles cut through the water, preying on any animal they can overpower. Their exclusively carnivorous larvae are known as water tigers, long, slender creatures with hollow, sickle-shaped mandibles for seizing and sucking their prey.

One of the main prey of water bugs and beetles is tadpoles, and there are countless numbers of them in rain puddles and temporary waterholes. There are no tailed amphibia in East Africa, but a great variety of frogs and toads manage to live in almost all habitats except waterless deserts and above the snow-line. Some live in permanently humid environments, in swamps, reedbeds or even hopping about on the forest floor. Others survive in semi-arid areas where there may be no rain for three-quarters of the year. When rain comes, every pond or puddle becomes alive with frogs or toads. They consume between them an astronomical number of insects, and are themselves consumed in vast numbers, especially in the tadpole stage, by all predatory aquatic insects, crabs, wading birds of every kind, kingfishers, snakes, terrapins and other frogs and toads.

The courtship of frogs and toads is usually a noisy affair. As soon as a male finds water he sits in it and "sings". Common toads croak, a deep, loud

Left: Fairy shrimps and water fleas pass the dry season as eggs, which hatch when the pool fills at the start of the rains. They are both very important links in the rain-puddle food chain, feeding by filtering single-celled algae, and themselves supporting many larger carnivorous puddle animals such as the tree frog tadpole shown here. Fairy shrimps swim on their backs, the females towing a big red and pink sac of eggs.

Tree frog tadpoles feed at first by rasping algae; later they become carnivorous and feed on dead fairy shrimps, other tadpoles and so on. The water tiger, larva of a big diving beetle, hatches as a voracious carnivore and remains so throughout its life.

Yellow-billed storks, spoonbills and great white egrets excitedly fish for spawning toads at sunset. The noisy calls of frogs and toads attract not only males and females of their own species but many amphibian-eating birds also, so spawning usually takes place at night when the birds are roosting. At the start of the rains, however, eager toads are often unable to refrain from calling until after dark.

Frogs and toads may have one large vocal sac beneath the throat or, as in this big groove-crowned bullfrog, two smaller lateral ones. Toads and reed frogs keep their balloons pumped up all the time, whether they are producing noise or resting between calls. The bullfrog's two small balloons only appear at the actual moment of sound production.

Tiny reed frogs blow up an enormous balloon when they are calling and produce astonishingly loud sounds for their size. When hundreds or thousands of individuals are calling in a swamp their ringing metallic chorus can be nearly deafening

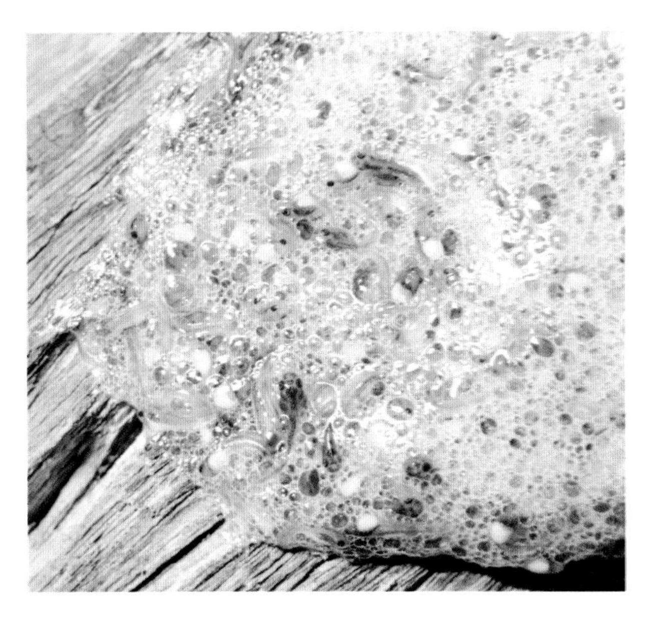

The grey tree frog spawns in vegetation above a rain pool, in clumps of grass or on a log, or high in the branches of an overhanging tree. As the eggs are laid, the pair beat the albumen into a froth with the hind legs to form a foam nest.

Above, right: Inside the froth nest, the grey tree frog's 150 white eggs hatch into tadpoles in 3-4 days. Two days later the tadpoles' wriggling liquefies the froth and softens the crust, and they drop into the water to complete their development in the usual way.

"Waaa-waaa" which attracts all others of that species within a wide radius. The chorus and splashing attract storks too, so spawning takes place mainly at night when the storks are roosting. A shower of rain stimulates a fine chorus from the tiny reedfrogs, too. Only 2 centimetres long, they are capable of producing astonishingly loud sounds for their size, and when hundreds of them are calling in the reeds their metallic tinkling chorus fills the air. Another characteristic frog noise of the rainy season is a melodious, watery "Boink", loud and penetrating from near to. The call is ventriloquial and seems to come from all over the place, from the water, from grass and bushes, even from up small trees. Any sudden noise stimulates a volley of "boinking". The sounds travel in bursts, and the makers are almost impossible to find. But the running frog, the maker of these calls, is common all over East Africa in open grassland from sea shore almost to snow-line. As its name implies, it seldom jumps, but runs or walks.

Frogs and toads lay large numbers of eggs, in some cases thousands, indicating that mortality is high. Many eggs must hatch in order for one or two young to reach maturity. A pair of common toads, for instance, may lay 24,000 eggs at a time, in strings of protective jelly. Grey tree frogs protect their eggs by laying them inside a foam nest. The female produces a thick mucus-like liquid, which she and her mate whip into a white froth with their hind legs. About 150 eggs are laid in this froth, which is always placed in vegetation over a waterhole. The outer layer of the froth hardens in the sun into a meringue-like crust. In three or four days the eggs hatch, the lower layers of froth liquefy as the larvae swim about, and finally the tadpoles drop into the water.

Three stages in the life history of the clawed frog: the unique catfish-like, colourless tadpole with barbels and tiny hind legs; an older tadpole with well-developed limbs and resorbing tail; and a tiny, newly-metamorphosed, tail-less froglet.

In contrast to most frogs and toads the courtship of the entirely aquatic clawed frog is an exceptionally quiet affair. The only call it makes is a

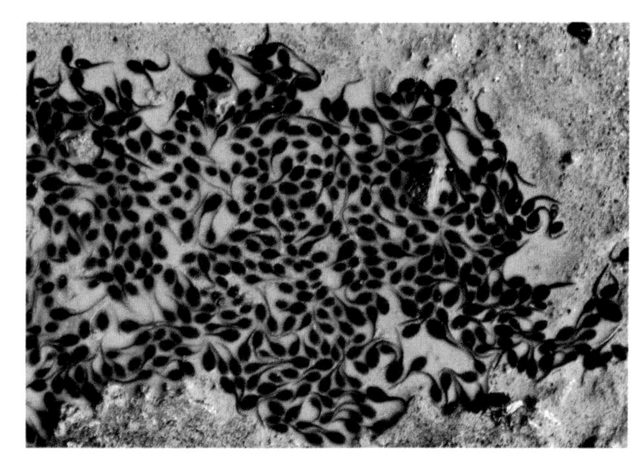

The striped pyxie toad lays its eggs in rain puddles. In some districts where the soil is very porous, the puddles may seep away and evaporate quickly. In shallow water the tadpoles gather in bands.

faint ratchet-like chirping underwater. Its tadpole in the legless stage looks like a small catfish because of the long slender barbel on each side of its wide mouth. Unlike other tadpoles it does not rasp algae but at first sucks in protozoan-rich mud and filters the excess water through gill-like slits on the sides of its head. The adult frog and older larvae have insatiable appetites, which make them of considerable ecological importance. They feed on any animal small enough to be raked into the mouth with their long sensitive fingers. The frogs are well camouflaged by their mottled colour, and their smooth skin makes them difficult to hold; but they are preyed upon by large fishes and herons in the lakes and by terrapins, kingfishers and storks in the puddles.

Frogs and toads are most abundant in the areas of higher rainfall where pools are likely to remain for at least three months, long enough for tadpoles to develop at a normal slow but steady rate. However, in certain areas of less reliable rainfall, particularly on porous volcanic soils, the striped pyxie toad is found, whose only nurseries may be ephemeral puddles which quickly evaporate. It has therefore acquired several remarkable adaptations. Firstly its eggs develop very rapidly, hatching in two or three days. Secondly, the tadpoles, instead of hanging for several more days developing mouths, are ready to start feeding the next day. If conditions are good, the tadpoles may double their bulk in the first day of feeding, and nearly double it again in the second day. After that growth is slightly less rapid, but the metamorphosed tadpoles are ready to leave their rapidly shrinking puddle in about three weeks.

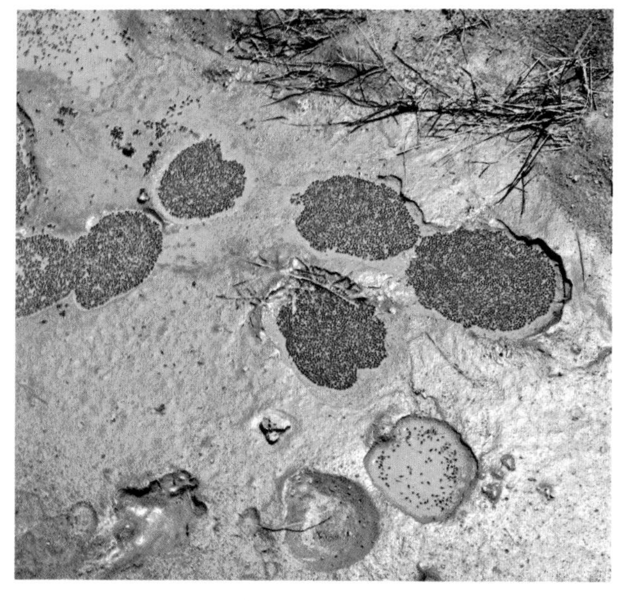

As the water recedes, tadpoles congregate in any depression, here in the hoof-prints of the antelope that drank up much of their already small puddle.

Even small footprints of wader and steinbuck become filled with tadpoles as the puddle evaporates faster than they can develop.

Finally, the layer of tadpoles bakes into a crisp pemmican biscuit in the hot sun. With the next rain it will be converted into rich food for the next crop of striped pyxie tadpoles.

Even such giant tortoises as this big adult African spurred tortoise disappear during the dry season, presumably hidden among boulders where their boulder-like carapaces escape notice. With the flowering of the thornbushes and the coming of the rains, they emerge from aestivation to feed and breed.

It often happens, however, that feeding conditions are poor, since there may not be much organic matter in temporary puddles. In this case the tadpoles band together in shoals and move slowly over the bottom, flailing with their tails to set up a current which will bring up all available food from the mud. At this stage many tadpoles die of overcrowding or are cannibalized, a biological necessity which ensures that others will develop faster. The banding together of the tadpoles also has the effect of deepening the part of the puddle in which they are congregated; what little water is left flows down into the depression, lessening the surface area and cutting down evaporation. Complete drying out of the puddle may be delayed sufficiently by this method for the tadpoles to finish their development. They also congregate in natural depressions, in the spoor of buffalo or antelope or in footprints of wading birds. But in really poor conditions the tadpoles are unable to complete their cycle in time. Thousands of them are stranded and the hot sun bakes them into a crisp sheet. But their lives have not been wasted. Their bodies become part of the material of the dry puddle bed, and the puddle that forms there in the next season's rains will be extremely rich in the most nutritious food for tadpoles—the bodies of dead tadpoles. Next year the tadpoles in this puddle will develop more rapidly on this diet and metamorphose at half the size so that their prospects of winning the race against evaporation are brighter than are those of tadpoles in puddles that did not dry up the year before.

Reptiles breeding

Rainfall governs the onset of breeding in the amphibia, for without rain those away from permanent water would have no pond or puddle nurseries. Rainfall is also the decisive factor governing the breeding of reptiles. In temperate climates where cold and short days inhibit breeding, warmth and longer days stimulate it. But in tropical regions, especially on the equator itself, where day length is constant and it is normally warm enough for breeding all the year round, the coming of the rains is the seasonal event that triggers it off.

Many kinds of reptiles are egg-layers—agamas, geckos, skinks, monitors, crocodiles, tortoises and terrapins. The ubiquitous side-necked helmeted water tortoise or terrapin is as mobile on land as in the water, and turns up in the merest rain puddle in the middle of dry bush, to feed on the tadpoles and insects. When danger threatens, it buries itself in the mud of the puddle bottom, sometimes heaving up and stranding tadpoles as it does so. All terrapins are carnivorous, but all tortoises are vegetarians. When the rainy season comes tortoises awake from aestivation to feast on seedlings and fresh green growth. But it is amazing that tortoises can still exist in much of their range: it appears barren even in the season of greatest growth, because the superabundant goats have stripped to bare twigs and earth all edible vegetation below goat-reach.

Tortoises and terrapins dig nests for their eggs in rain-softened ground. They also urinate copiously while digging, wetting the ground further. For them it is the actual softening of the ground by rain that is crucial in the onset of breeding, as is also the case with the Nile crocodile. After the start of the

A baby Nile crocodile surfaces among floating water fern.

The male three-horned chameleon has a prehistoric look, but with his grasping feet, prehensile tail, leaf-like shape and colour, and rocking, wind-blown-leaf gait he is wonderfully well adapted for living in bushes. His tongue can accurately swot a fly as far away from him as his own length.

first rains, crocodiles mate and the female digs her nest in a shady rain-softened, sloping shore. The incubation period coincides with the following short dry season, during which the mother guards her nest against monitor lizards and other predators. Soon after the onset of the long rains the eggs begin to hatch. The baby crocodiles are very vocal, like baby birds, and their underground chirpings prompt the female to release them by digging up the nest. She goes on guarding her fifty or so hatchlings for at least six weeks. When she basks in the shallows at the edge of the lake the young climb out of the water onto her back and head. Periodically she submerges, leaving the babies swimming. When she resurfaces they climb back onto her. But in spite of her care there is very heavy mortality among the young crocodiles from monitors, marabous, eagles and other crocodiles.

Like chelonians and crocodiles the majority of lizards are egg-layers. Geckos cement their eggs in small clutches under bark or stones, skinks lay them under rocks, agamas may dig nest-holes in the earth. Most chameleons, too, are egg-layers and dig nests in the ground: the flap-necked chameleon excavates a hole with her hind feet as deep as she can reach while holding onto the rim. She lays thirty or forty eggs, and afterwards tamps the earth down and even scatters dry twigs and grasses over the nest to hide it. Some chameleons avoid the perils of descent onto the ground by giving birth to their young in the bushes. In East Africa chameleons found up mountains are viviparous. The two-lined or high-casqued chameleon is found at altitudes at least up to 3,000 metres on Mount Elgon, the Aberdares and other mountains, clinging to lichen-covered St. John's wort bushes, or among everlasting flowers. Lower down, at about 2,000 metres Jackson's chameleon is very common in places. The three-horned males look like miniature *Triceratops* dinosaurs, and in territorial disputes lock horns in a slow-motion trial of strength; the females are hornless. The advantage of retaining the developing young within the body in reptiles that live at high altitudes is that the female, by basking in the morning sun, or sheltering from too much heat at midday, can select the most comfortable temperature for herself, which is probably also the optimum one for the eggs.

Viviparous chameleons are delivered in a clear jelly membrane from which they quickly struggle free. They soon move off through the acacia foliage to lead independent lives.

The embryos are protected from desiccation or too much damp, and from bacteria and fungi, in a superbly camouflaged mobile incubator. The babies are born encased in a jelly membrane which sticks to twigs or leaves. They quickly struggle free of this wrapping, and set off with eyes swivelling and tiny hands clutching, perfect miniature replicas of the mother.

Among snakes, as among lizards, the majority are egg-layers. Some even show maternal care: the python and blind burrowing snakes incubate their eggs, remaining coiled around them for some weeks until they hatch. Some baby snakes may be mainly insect-eaters until they have reached a large enough size to tackle the normal prey of their species. The egg-eating snake, however, need not eat for several months after hatching and its first meal will probably be a gecko's or other small lizard's egg. As it grows it will climb trees to take the eggs of weavers and finches. Baby egg-eating snakes are, therefore, most likely to hatch before the breeding season for other reptiles and birds.

A newly-hatched rhombic egg-eating snake among the cast wings of termites. Egg-eaters are non-venomous but in certain areas closely mimic the colour pattern of the highly poisonous night adder. In semi-desert areas where the night adder does not occur, the egg-eater appears to mimic other venomous snakes. It is not certain whether this is true mimicry, or convergent evolution: both species may have arrived at a similar patterning and coloration as the best camouflage for the area in which they live.

Nesting birds

Rainfall is the dominant factor in stimulating breeding in birds. For some, the sight of falling rain is enough, but insectivorous species wait until new foliage has brought an abundance of insects. Weavers, which need long grass for weaving their nests, wait until it has grown, but swallows, which use mud, can start building soon after rain. Where there are two rainy seasons in the year some small birds may breed in each.

The onset of the rains, therefore, is the signal for a great burst of breeding activity. The arid bush country is refreshed and springlike with new greenery and flowers, and every thorntree or thicket provides a nest site for at least one pair of birds.

The bush is full of bird noises, some harsh, some melodious and sweet. Among the most characteristic sounds are the songs of the duettists. Duetting is restricted to various tropical groups, and is unknown in temperate regions where female birds usually do not sing. In duetting species, both cock and hen sing. Their notes are different, not sung at random but alternated antiphonally in such a precise way that the calls sound as if they could only have come from one bird. Most duettists live in dense vegetation, and the calls help to keep the pair in contact.

The best known duettists are the bou-bou or bell shrikes. Some have a simple duet of only two calls. The black-headed gonolek, with black back and brilliant red breast, skulks in thick bushes, the male calling a short whistling "Yoik" which is answered by a tearing-cloth sound from the female. In contrast the black-and-white tropical bou-bou has an extensive repertoire of antiphonal melodies, some tonally very pure and pleasing but varying from district to district. The cock may give a series of bell-like calls, the female answers with a harsher note. A pair rehearse their duets and together evolve new patterns which they do not share with other pairs. This means that if a cock calls and another bird responds correctly, she is his mate.

Some other bird groups that duet in pairs do so as conspicuously as possible, perched close together on top of a bush or a dead tree. Red and yellow barbets shout "Tweedle-de-tweedle" over and over again, while d'Arnaud's barbets sing a loud four note song perched side by side, one with tail cocked up, the other beak up, tail down. Pairs of red-billed hornbills call a monotonous and continuous "Wot, wot, wot". As the tempo increases, so their wings half open until at the end of the duet they look like burlesques of heraldic eagles. Fish eagles also call conspicuously, flinging their heads up and back, a flashing white visual signal to augment the sound.

The red-backed scrub robin delays its nesting and therefore the hatching of its chicks until insects are plentiful, a little later on in the rains. Both parents bring caterpillars, flies and small grasshoppers for their nestlings.

Day-old red-backed scrub robin nestlings are naked and helpless, but have remarkably strong necks to hold up their big heads. In contrast to the nestlings of most other birds, which gape in response to vibration, they gape vertically as soon as they hear the musical call of a parent bringing food.

A crowned plover chick rests, half out of the egg in an uncomfortable-looking nest, a mere scrape in the ground lined with fragments of volcanic rock. Hatching may take all day, though it can be much more rapid. In another hour this chick was out of the egg with down dried, already leaving the nest to follow its parents.

A crowned plover settles itself to brood newly-hatched chicks until their damp down has dried. While the chicks are small, the parents find food for them and call them to it, indicating it by pointing with the beak. Even when full grown and independent, the chicks may stay with the parent.

The greater flamingos' nuptial display is a slow-motion ballet in which each bird walks with neck full length, turning its head smartly first to one side, then the other. Suddenly several birds together snap their wings open in salute.

D'Arnaud's barbets calling together. Duetting strengthens the pair bond; during the breeding season arid bush country rings with the noisy, jubilant calls of barbets.

The cock kori or giant bustard displays by puffing out the feathers of neck and under-tail, a conspicuous visual signal as he tours his territory on the grassy plain in the early morning.

Their loud ringing call of three descending notes is possibly the most evocative bird sound of Africa.

Conspicuous and evocative bird calls of other areas are the loud croakings of touracos in the forests; the various mournful or plaintive cuckoo calls which are said to herald rain; the distinctive laughing or lamenting of doves in early morning; the bubbling of the coucal; the cackling of francolin or spurfowl in the bush; and the loud "Er-anna, er-anna" of the white-bellied bustard of the plains. In these the songs are more conspicuous than the birds' displays, but in others it is the displays that catch the attention. The greater flamingo gives a magnificent, majestic, wings-out display; on a diminutive scale it is echoed by the grey-headed kingfisher which simultaneously trills penetratingly. The kori bustard tours his territory with a stately walk, his tail cocked up to display the conspicuous white of fluffed under-tail coverts. Crowned cranes have an excited but stately dance which they perform while melodiously calling with a plaintive yet haunting sound. Lesser flamingos perform a parade in which a hundred birds march closely-packed with heads up and necks flushing bright pink, long red legs flashing, voices cackling and honking in communal excitement.

In some birds with particularly eye-catching displays the males have elaborate ceremonial plumages for the nuptial season. The bishops are glossy black with much brilliant red or yellow. Whydahs are not so colourful but have superb long tails. A bishop male displays over his territory, flying up with a peculiar undulating, floating flight, rump feathers puffed and giving aggressive sizzling calls. He may have half a dozen sparrowy females nesting in his territory. Jackson's whydah makes circular rings in the grass from which he displays. Some species spring into the air and fall back into the grass repeatedly, giving an extraordinary impression of ragged black objects being regularly tossed up. No whydah nest has ever been found; all species are thought to be brood parasites, some in the nests of waxbills. Female whydahs watch their hosts' nesting preparations

Lesser flamingos bunch close together with heads high and neck feathers fluffed. They parade along in an excited, noisy communal display, birds on the fringe occasionally saluting with open wings.

Little weavers build their nests overhanging water. In spite of the fact that these nests are sited at the very tips of twigs, agile climbing snakes, such as mambas and egg-eaters, usually have no difficulty in entering them to take eggs or chicks.

The male red-headed weaver uses pliant green twigs to weave his nest; other species may use grass blades. This bird has almost completed the foundation ring on which the globular nest will next be built up.

and the sight causes them to come into breeding condition. They lay two or three eggs in the waxbills' nest, but unlike baby cuckoos, baby whydahs do not evict the host's own eggs or nestlings. The colour of the whydah's eggs, the mouth markings of the nestlings, and the juvenile plumages, mimic those of the specific host, and fledgling whydahs stay with their foster-family for some time after leaving the nest. The cocks pick up their foster-father's vocabulary and later use it in their own courtship.

Other birds with special nuptial plumes include three nightjars which have splendid elongations of wing or tail feathers. The long-tailed nightjar has elongated central tail feathers; the pennant-winged has very long primaries, almost twice as long as itself; but most extraordinary is the standard-winged nightjar which has in each wing a very long-shafted feather with a broad flag at the end, and in flight looks like one large bird followed by two tiny ones. In the breeding season the wattled starling moults its head feathers and grows extraordinary black wattles on crown and throat while the rest of the skin of the head becomes bright yellow. Some sunbird males moult into brilliant metallic plumages for the breeding season.

Nesting weavers are very conspicuous, the cocks brightly coloured in the breeding season with yellow or red breast off-set by black bib, cap or mask. The male builds elaborate hanging nests, beautifully woven of vegetable fibres, often grasses. He starts by constructing a ring at the tip of a branch, and from this hangs upside down to advertise for a mate by wing-flapping and calling in urgent buzz or chirping chatter. The nest is built up around the ring, a tough sphere with a side entrance, protected by porch or pendant tunnel. Weavers are polygamous, and may build nest after nest. Their colonies are a seethe of excitement as the males flutter upside down, displaying yellow or red breasts or underwings to the sparrowy females. Other varieties of weavers build huge communal nests that look like hay stacks.

Greater flamingos' colonies are also the scene of intense excitement as nesting gets under way on some rocky island. Often at this time black clouds build up for the afternoon downpour, and winds whip up banks of spume along the shores. The flamingos on one island all lay almost simultaneously, and the young hatch within a week of

Egg-laying is remarkably simultaneous in a greater flamingo nesting colony, so, except where rising water has caused the odd egg to be abandoned, the chicks all hatch within a few days. Newly-hatched chicks have pink legs and pale silvery down; as they grow, down and skin darken. Nests are so close together on the island that the adults are constantly squabbling to maintain individual distance.

In some years the entire crop of baby greater flamingos is eaten by marabous within a week or so of hatching. The adults bravely attempt to defend their chicks, but are no match for the storks.

A marabou grabs a chick and swallows it whole. Predation is generally seen as a natural means of regulating the numbers of the prey species and keeping the stock healthy by culling the weak. In this case, where perhaps only a dozen chicks in a thousand survive, it is difficult to see how the flamingo stock benefits from the attentions of the marabous.

each other. But within days of the first chick's appearance, marabous come planing down onto the island and begin gulping down the hatching chicks, in spite of brave protests from the parents. As one marabou becomes gorged, another takes its place, and between them the whole island may be cleared of flamingos in a very short time. Only a few young may be shepherded by their parents to islets in deeper water, and so escape

Migrants

In much of East Africa the long rains occur during the Palearctic winter, so that the tropical season of abundance coincides with the temperate season of scarcity. Many European birds therefore migrate to Africa in the autumn, about 600 million arriving by way of the Mediterranean and trans-Sahara route. Harriers and eagles are funnelled down the Rift Valley, where they remain to feed mainly on rodents. Large eagles are grounded by rain. Just before a storm they perch on a small tree or fence post, and sit hunched and sodden-looking until after the rain has passed. But swifts and hobbies follow the storms; when black clouds build up they appear by the hundred, hawking flying insects brought out by rain. European swallows appear in huge flocks, and are very conspicuous when they first arrive, clustered in thorntrees, particularly near the soda lakes where there are abundant lakeflies for them to feed on. Vast flocks of waders also descend on the Rift Valley lakes. Sometimes migrant ducks, particularly hordes of shovelers from Europe, African pochard, Hottentot teal and Cape wigeon, congregate in rafts in uncountable numbers.

A local migrant that lives in a world of perpetual abundance is the glossy-black Abdim's stork. It breeds in the northern savannahs during the rainy season there, then moves southwards, appearing in East African grasslands with the November rains. During the next dry season it continues south with the rain belt then returns north with it and reappears in East Africa with the April rains. Thus it spends its whole life in the grasslands when these are wet and abundantly swarming with insects.

At one time Abdim's storks were able to feast on the swarms of locusts that also followed the rains. The migration pattern of locusts is particularly suited to the climatic regime of East Africa. They

Six hundred million migrant birds arrive in Africa each Palaearctic autumn, many of them waders which descend in great flocks onto the Rift Valley lakes. Here a flight of ruffs alights among the reeds, where a female Bohor reedbuck is feeding and a flock of African spoonbills resting.

fly downwind in swarms towards frontal systems of converging air flows and so accumulate in the rainy areas. There, the flush of new vegetation provides good feed for the next generation of hoppers, who march along in bands, devouring every green thing. The migrating swarms used to do catastrophic damage to crops and all green vegetation, and were so numerous that although storks and other big insectivorous birds gorged themselves on them, they had no effect on the insects' destructive powers. Since 1944, however, the plagues have been kept under control by anti-locust organizations which prevent them building up in the outbreak areas. The multi-millions are now rarely seen in East Africa, only solitary individuals.

Other insects dependent on seasonal vegetation also follow the rains. The African migrant butterfly sometimes migrates on a locust-sized scale. Millions of white males flutter past like a snowstorm, a few sulphurous females among them. A small, nocturnal, brown moth, *Spodoptera exempta*, whose larvae are known as army-worms, migrate, flying with the wind, unaffected by the coolness of highland nights or the falling of rain. The females lay their eggs among freshly sprouting grass. When the caterpillars

Thousands of European shovelers winter on the East African lakes every year. Drakes are often in partial eclipse plumage when they arrive, but most have moulted into their distinctive breeding colours before they depart in the spring.

Army-worm caterpillars rival locusts in their destructive powers in grassland. During an army-worm invasion, all insect-eating predators, such as this small skink, gorge themselves on the glut but have little effect in stemming the tide of munching caterpillars.

A royal pair of termites sets off in tandem after a very brief flight. The female has shed both pairs of wings, but the male still has a pair attached. They cannot have flown far before pairing, for the attendant workers and two castes of soldiers are still milling about.

hatch, they eat their way across grassland in a living green-black carpet. Birds descend on the feast: storks, ravens, hornbills, as well as smaller insect-eaters, devour vast numbers of caterpillars, but there is such a glut of them that millions more remain. They graze good pasture to stubble and in more arid places do not even leave the stalks. Suddenly they all disappear underground to pupate. A further rainfall stimulates the emergence of the next crop of moths: forty-five million can take flight in a single night.

Insects

Another spectacular sight of the rainy season is the swarming of the termites. For many weeks winged sexuals have been waiting underground for the right conditions for their maiden and only flight. The rise in humidity with rain brings the right conditions. Over a wide area, all the nests of a given species swarm simultaneously, so that many millions of insects emerge together. The direction of flight is quite random and unless there is a strong wind to carry them the insects do not fly far. There is no attempt to mate in the air, so the flight is dispersal rather than nuptial. When the termites land, their wings easily break off at their bases. Now wingless, they run about searching for a mate, and then proceed to tandem, the male closely following the female, stroking her abdomen with his antennae. The pair search for a suitable cranny in which they can excavate the beginnings of a new termitarium. Only when this is done do they copulate.

Flying termites are supplied with fat and proteins so they can start new colonies without having to leave the shelter of the nest to forage. They are therefore very valuable as food for other animals and a great many creatures gorge themselves on them whenever they can. During a daytime emergence great vultures and tawny eagles toddle incongruously after them on the ground, with hornbills, storks, baboons and monkeys; while chanting goshawks, grasshopper buzzards, pygmy falcons,

Several pairs of royal termites may excavate a nest together in which they will found a new colony. When safely underground they will mate and the female will begin egg-laying.

Right: A termite–ant battle: As winged termites emerge for their dispersal flight, a huge soldier ant seizes a worker termite in its powerful jaws. It is itself immobilized by termites holding its legs, while a worker ant in turn fastens onto one of the termites.

rollers, bee-eaters, shrikes, barbets, glossy starlings and others hawk them from nearby trees. At night Verreaux's eagle owls, pearl-spotted owlets, nightjars, and bats take the places of the diurnal animals. When larger predators are satisfied the lesser ones feast: geckos, toads, crabs, scorpions, rats, hedgehogs, shrews, and bushbabies by night; dwarf mongooses, skinks, elephant shrews, spiders, frogs, and ants by day. As termites swarm irregularly no animal can be entirely dependent on them, except perhaps the hobby, which like Abdim's stork moves south with the rain and then returns with it, feeding largely on flying termites on the way.

Ants also swarm at the start of the rains. Theirs is a nuptial flight, for they copulate on the wing. But the male takes no further part in the foundation of a new colony as he dies shortly afterwards. When she lands the female bites or shoulders off her wings and searches for a place to start a nest. Not all species of ants produce mating swarms; queen safari ants for instance, are wingless, though the males are those large brown wasp-like but harmless creatures known as sausage flies that bumble and zoom about lighted rooms at night during the rains. It is not known how the males find the

All sorts of insect-eating animals feast on the nutritious flying termites. European storks, winter visitors to Africa, are among the birds that take termites whenever there is an emergence.

When conditions are right for a termite emergence, they are often right for flying ants also. With ants it is a true nuptial flight, as mating takes place on the wing.

Insects are protected from predation from birds, lizards and other animals in a variety of ways. Some gain protection from their cryptic coloration, by remaining motionless among plant parts of a similar colour. The bark mantis is almost invisible against bark, while the leaf mantis it is eating, obvious in this setting, would be beautifully camouflaged among green leaves. Other mantises mimic flowers and grass.

Red, yellow or orange combined with a dark colour advertise that an insect is poisonous, and birds quickly learn to ignore one flaunting these colours. The lubber grasshopper, unlike most of its family, is slow, clumsy and conspicuous. It is extremely unpalatable, exuding a caustic yellow fluid from joints in legs and thorax when disturbed. It enhances its warning by flashing its wings open, displaying large areas of colour and abdominal stripes. Among bees and wasps, bold stripes are common.

queens. Sausage flies and other flying ants are pounced on during the glut by every insect-eater, habitual or opportunist.

Enormous population explosions of other insects occur during the rains. There are insects everywhere, in nearly every environment and exploiting almost every kind of food, though as with mammals, green vegetation forms their major food source. Almost all butterfly and moth caterpillars, all grasshoppers and most crickets, many flies and beetles, are leaf-eaters. All bugs except a few blood-suckers suck plant juices. So with rain-stimulated plant growth there occurs a great proliferation of insects, whose ecological importance is enormous since they convert a good percentage of all plants into animal matter.

Insects are not only conspicuous through force of numbers, some are conspicuous through the sheer size of the individual. Nearly all exceptionally big insects are tropical because only in a warm climate can long respiratory systems work efficiently. Some very large insects occur in East Africa: swallowtail butterflies in the forests; huge saturnid moths in the bush; and very big scarabs or dung beetles.

Dung beetles emerge from their underground cocoons at the beginning of the rains, and soon set to work burying dung. They can be seen homing onto fresh dung almost as soon as it hits the ground. They fashion the dung into balls much larger and heavier than themselves, and trundle them along by running backwards rotating the dung-ball with their hind legs. When a suitable spot of soft ground is reached, each beetle buries itself and its dung-ball in an underground chamber. There the beetle shreds the dung, remoulds it into a beautiful pear-shape or oval, and lays an egg in a hole in the top of it. The dung remains moist underground, and the larva, when it hatches, eats away the inside of the pear and then pupates inside the remaining thin outer shell.

Some dung beetles are compulsive dung-buriers and bury all they can find. Individuals of a small species less than 2 centimetres long can bury 100 cubic centimetres of dung each in a night. The work of dung beetles is therefore of great importance in the ecology of a region. Underground the dung is acted on by bacteria and transformed into substances useful to plants; left above ground it dries and remains useless. By burying dung containing seeds (page 117), the beetles assist their germination. In addition, their work is of prime sanitary importance, since many diseases are transmitted by faecal matter and faeces-visiting flies.

An African monarch or milkweed butterfly ostentatiously spreads itself before its maiden flight, displaying bright colours that warn of its poisonous nature. The female of the diadem, an unrelated and innocuous butterfly, mimics the monarch and so is protected from predation, but the diadem male is an unprotective velvety-black with a round, white, violet-edged patch on each wing.

A huge saturnid moth with a wingspan of 15 centimetres also spreads himself before his first flight. He is not poisonous; while at rest during the day his cryptically-coloured forewings obscure the hind wings. But if alarmed he flashes open, displaying the very striking eye-spots. The sudden sight of such brilliant, owl-like eyes, works well as a deterrent to insectivorous birds.

Dung beetles are of great ecological importance, for they are compulsive buriers of dung. Females trundle balls of it backwards until they find a suitable spot of soft ground in which to bury it.

A close-up of the inside of a *Hydnora* beetle-trap flower. Small dung beetles, attracted by the smell, have tumbled down inside and cannot climb out again up the waxy, overhanging petals.

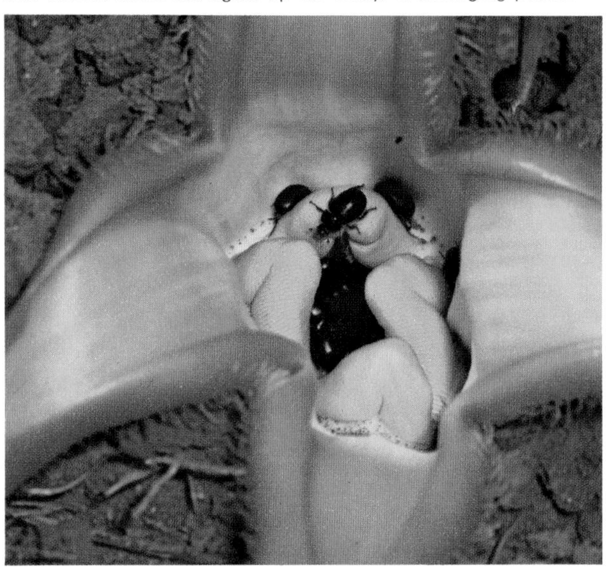

The very biggest dung beetles, which are stout, well-armoured, almost cubic creatures about 7 centimetres long, would seem to be large and tough enough to be left alone, but they are eaten by sacred ibises which dig them up from their underground burial-chambers and swallow them whole, and by lilac-breasted rollers which beat them to a pulp on an anvil-stone first. Augur buzzards and pale chanting goshawks strip off the wings and elytra and swallow only the abdomen, leaving the head and thorax to tramp steadily on as if nothing had happened. Smaller dung beetles are the staple diet of banded mongooses, who follow in the wake of feeding herds of elephants and buffaloes and rake over their dung to find the beetles. There is even a plant, *Hydnora*, that "eats" small dung beetles. A parasite on the roots of acacias, its presence is unseen until huge red beetle-trap flowers burst open at the surface of the ground. Dozens of beetles are attracted by the flowers' foetid scent, crawl on the smooth waxy petals and tumble down inside. They are unable to climb out of the slippery trap and quickly perish, to be digested as the flower wilts and disappears below ground again.

In the forests, butterflies may be on the wing at any time: gorgeous *Charaxes*, orange, iridescent blue, mother of pearl, bask on sunlit foliage in the glades or speed among the trees with powerful, direct flight. But at this season even arid bush comes alive with butterflies: big yellow swallowtails, orange-brown monarchs, small sulphurs that cluster in brilliant patches on damp mud, magenta-tipped whites and orange-tipped lemons, tiny long-tailed blues, peacock-eyed pansies. In the hot sun, even the smaller ones fly much faster than the butterflies of temperate regions. When the sun goes behind a cloud, they touch down and hide among foliage to wait for the shadow to pass. In cooler cloudy weather butterflies at a salt-lick close up when the sun goes in; when it comes out they open wide and bask.

One of the most interesting among the many remarkable butterflies of East Africa is the gaudy commodore, which parallels the great contrast

Citrus swallowtails and white butterflies sucking up the dissolved mineral salts in wet mud beside a wallow, probably where a big animal has urinated. Some whites, like other African butterflies, vary with the season. Dry season forms are paler than wet season forms, with reduced dark markings.

between the wet and the dry seasons by having distinct seasonal forms. In the dry season the butterfly is reddish-orange and brown to match the predominance of these colours in the dry season landscape. In the rains it is predominantly blue, chequered with black and red. Intermediate forms also occur, purple-blue with a broad red band, and both wet and dry season forms may be seen on the wing at the beginning of the seasons.

There are so many plant-eating insects that, if they are not completely to dominate the area, they must be controlled. Spiders are a check to some extent on insects. They are an immensely varied order, second only to the insects themselves, and because they are all carnivorous their effect on other invertebrates is profound. Very many kinds of spider use their silk to weave elaborate webs in which to snare insects many times their own size. Others, such as baboon and wolf spiders, live in burrows in the ground from which they emerge at night to leap on their prey, as the tiny jumping spiders do by day. Yet others, such as crab spiders, lie in wait, often beautifully camouflaged in a flower, hidden from their own enemies and from their prey.

But probably the greatest natural check on the hordes of plant-eating insects are the parasitic wasps. The females tirelessly search for individuals of the particular insects which are hosts to their species. When a female wasp finds a host she pierces it with her ovipositor and lays one or many eggs inside its body. When the larva or larvae hatch they feed within the host, first on non-vital tissues, so that the host appears to develop normally. Later they attack the vital organs, then emerge from the empty skin, spin a cocoon and pupate.

Many parasitic wasps are extremely small, among the smallest known insects. Their larvae develop within a single egg of the host, or within a tiny host such as an aphid. Some species are counterchecks on other invertebrate predators, and may be very large. The spider-wasps are black or metallic blue insects up to three inches long, with long legs and iridescent blue wings. Their prey is almost exclusively spiders; the very big species

The tiny yellow jumping spider is diurnal, and stalks its prey by stealth, catching it by leaping onto it. Relatively huge prey such as bluebottle flies, are quickly subdued by injections of paralyzing venom from the spider's fangs.

The big wolf spider lives in a burrow lined with silk, and emerges at night to run actively after its prey. The female carries her eggs in a cocoon attached to her spinnerets. When the baby spiders hatch they swarm all over her body and she has to keep them off her face and headlamp-like eyes by frequent windscreen-wiper movements of her pedipalps.

The crab spider can assume the exact coloration of the background on which it is sitting, here a cluster of milkweed flowers. It sits motionless with front legs held wide, waiting for an unsuspecting insect, here a small blue butterfly, to alight.

The caterpillar of the African monarch butterfly feeds on milkweed, a poisonous plant. It stores heart poisons in its body tissues, and if swallowed by birds, causes severe vomiting and distress. It advertises its poisonous nature by its black and yellow stripes, yet it is parasitized by tiny wasps. The female wasp lays her eggs in the caterpillar; when the grubs hatch they feed on its tissues and then pupate around its empty skin.

Spiders are a check on the insects; spider-wasps a counter-check on the spiders. This large black spider-wasp is dragging a huge baboon spider it has stung and paralyzed. The spider will provide an immobilized, living larder for the wasp's larvae.

prey on baboon spiders, which are stung so accurately in the main nerve centre they remain alive, though paralyzed, for forty days, providing fresh food for the wasps' larvae. Large spider-wasps bury baboon spiders in the ground; smaller spider-wasps are potters, making vases of cemented mud which they stock with stung spiders for their larvae to feed on. Other potter wasps provision their beautiful little pots with paralyzed caterpillars. Some parasitic wasps reproduce by parthenogenesis; in others a single egg develops into a multitude of individuals. The phenomenon of parasitism is immensely complicated, and is made even more so by the fact that many of the parasitic wasps are themselves the specific hosts to other parasitic wasps. For example the velvet ant (actually a wasp whose flightless, black and white females run about on the ground and can give a very painful sting) parasitizes another species of wasp (striped blue, black and yellow) which lives in scattered colonies on sandy, sloping shores of soda lakes and stocks its burrows with blue-bottle flies.

Other stinging, biting and parasitic invertebrates also proliferate during the rains, to the discomfiture of many animals, from tortoises to lions and elephants. Mites and ticks abound. After rain, hungry ticks—larvae, nymphs or adults—can be seen on the heads of tall grasses waiting to climb onto any large animal that brushes past. Ticks can survive very long periods away from their hosts, even in arid places. Some show almost incredible powers of water conservation, and can live for ten years or more without food or drink. Whilst on their host they gorge themselves on its blood, then drop off to the ground again. One female may lay 4 to 8 thousand eggs, which probably implies the chances of finding a host are slim. Once attached to the host, ticks are very difficult to remove, and cannot easily be dislodged by scratching or nibbling. Primates, with their nimble fingers, are uniquely able to rid themselves and each other of such parasites. Large animals rely on tickbirds.

Tickbirds or oxpeckers are the sole exploiters of a unique ecological niche. There are two species, the yellow-billed and the slightly smaller red-billed. The ranges of the two overlap, and they are not clearly separated ecologically; mixed flocks can sometimes be seen feeding together on the same animals. They visit a wide variety of game from rhinoceroses, buffaloes and giraffes down through zebras and antelopes to warthogs, and man's

Buffaloes resting at midday tolerate the red-billed oxpeckers that work over their hides, removing engorged ticks and wiping the blood off their beaks on the buffaloes' great horns.

Parasites proliferate during the rains. Blood-sucking invertebrates can be seen gorging on the throat and thighs of this Masai cock ostrich as he grazes on red oat grass.

Antelopes are much troubled by the swarms of biting, blood-sucking flies—*Stomoxys*, tsetses, horse flies—during the rains. This kongoni bull has been deliberately kneeling down to rub his horns and face in a mud puddle—an action which probably has sexual significance. After scratching his flanks with his horns, he twists round to gnaw an itch on his rump.

domestic stock. Some pastoral people welcome them, but today oxpeckers have disappeared from many areas where modern farming techniques involve the regular use of cattle dips. When ticks disappear from cattle, and game is no longer abundant, the tickbirds also disappear. They obtain the whole of their sustenance from the hides of their hosts, mostly by picking off engorged ticks with their specially-adapted flattened beaks. It is not so much the ticks themselves which provide the birds' food, as the blood which they contain. The diet is supplemented by blood-sucking *Stomoxys* flies, and by the blood and tissue taken direct from wounds on the living animal. These are consequently often prevented from healing for a long time, and even enlarged. Surprisingly, oxpeckers are tolerated remarkably by their hosts, even when they peck at raw wounds. Sometimes such wounds, especially those on the shoulders or flanks of rhinoceroses, are infected with microscopic parasitic worms (*filaria*). The oxpeckers may be disinfecting the wounds to some extent rather than feeding on healthy tissue; the worms prevent healing rather than the oxpeckers. Often their attentions are actively solicited, as when an antelope stands stock still with its ear presented for deticking. The tolerant mammals derive further benefits from the symbiotic relationship since the tickbirds warn them of danger; and in return the birds, as well as finding food, warmth and a mobile display ground, are provided with readily-obtainable hair with which to line their nests.

The tiny lilies that flowered at the start of the rains now bear big, succulent fruits. The leaves always appear cropped, probably by antelopes, when they first push out of the ground and after the flowers have wilted. The stem collapses under the weight of the ripening fruits.

By the end of the rainy season young roan antelopes are up to the withers in a sea of tall, seeding grassheads.

The end of the rains

At the end of the rainy season the showers become fewer and lighter, separated by longer rainless periods. Antelopes may be up to the withers in a sea of tall grasses above whose seeding heads the growing calves are hardly visible. At this time greater kudus give birth, and hide their new calves in the dense vegetation. Many plants have seeded, and bushes and trees are now about to lose their leaves to conserve moisture in the coming dry season. Some take on brilliant reds and yellows before leaf-fall, like trees in temperate regions in autumn. Small birds may still be feeding their fledglings, possibly the young of a second brood. Those birds that nest in the dry season are making their scrapes or collecting nesting material now. Others, which have finished breeding, go into post-nuptial moult: weavers, bishops and whydahs into a sparrowy plumage; nightjars, wattled starlings and some sunbirds into plumage almost indistinguishable from that of their females. The European migrants gather in flocks before they depart. Lakeside trees are loaded with swallows, European storks spiral in the thermals in flocks of five hundred or more before setting off for their breeding grounds. Butterflies feed on late flowers, grasshoppers munch drying grassblades. In the higher rainfall areas, all is lushness and late summer abundance, but in arid country the grass has already turned to standing hay, bushes and shrubs have dropped their leaves and wear the bare winter look typical of the dry season. Small animals are hidden underground, rivers are low and water holes drying: the country begins its next long season of drought.

An oribi and her kid at the end of the rains stand almost hidden in the tall grass. The broad-leaved trees are in full dark green leaf, but there are still a few flowers on one of the *Erythrina*, which blossom when the twigs are bare.

The flood plains of the Zambezi are backed by mopane forest in which greater kudu live. Mopane leaves turn red at the end of the rains and remain attached to the trees for a long time into the dry season.

European swallows cluster on lakeside trees before setting out on their return journey.

The African lion, the superpredator and lord of the plains, calmly surveys his domain. Yet he could face starvation if he is not able to hunt and feed without constant surveillance from tourists.

Epilogue

Almost nothing can be written about wildlife today without sounding a note of doom, for the dangers that confront ecological systems and the wild animals in them are many and real. Pollution of the environment is still only an emerging problem in East Africa, but the greatest threat is the rapid growth of its human population and the very limited nature of its resources. East Africa has great open stretches of land, but because of its extreme climate most of this is not good arable land, while primitive farming methods have already exhausted much of the soil. Land pressure is growing steadily as towns expand and farms spread. In the most fertile areas big ranches that produced food for the many are broken up into subsistence plots for resettlement by the few. In less fertile areas, people, cattle and maize spread at the expense of wild animals, trees and grass. The wild animals have been squeezed into East Africa's sixteen national parks and into a score of smaller game reserves. The national parks cover an area of about 32,000 square kilometres, but this is only three per cent of the land area of Kenya, Tanzania and Uganda combined. And even vital parts of the national parks and forest reserves are liable to be handed over to the people for settlement. As human populations build up around and even within the parks, the menace of poaching increases, free movement of animals on great traditional migration routes is impeded, and possibly the entire ecology endangered.

However, the revenue from wildlife and the interest shown in it by visiting tourists have convinced many governments that the animals and wildernesses must be preserved, and that a living antelope or zebra is worth more to the country as a tourist attraction than its carcass is worth as meat. The half-million tourists who visit East Africa's parks each year stimulate local economies and bring in much-needed foreign currency. But conservationists are now worried that the flood of visitors is having a harmful effect on the animals, particularly in some of the most popular parks. The great cats, especially, are subjected to nearly continuous observation by tourists, and may spend much of their daylight hours in the centre of a ring of motor vehicles. Cars may chase alongside them when they are trying to hunt, and they may even be prevented from making their kill or from eating it. The cumulative effect of visitor impact may be that lions, cheetahs and leopards are unable to get enough to eat. Similarly, game viewing too close to waterholes, particularly in the dry season, may prevent game from getting enough to drink. In the breeding season, the eggs of ground-nesting birds are at high risk on mudflats and short-grass swards, particularly in the vicinity of the spectacular and much visited soda lakes. Even the vegetation suffers from visitor impact. It only needs two or three cars to follow in the track of another across a piece of grassland for the wheels to leave permanent scars on the pasture.

The management of wildlife in East Africa, as anywhere else in the world, begins with the management of people, both indigenous and visiting. The obligation of the African governments to produce economic prosperity for their peoples and more land for human settlements must be balanced by plans for land-use that are more scientifically determined, with more thought for the morrow. There is a growing awareness of the value of wildlife in East Africa so it is very much to be hoped that governments will adopt far-sighted plans already proposed to leave major parts of ecological systems undisturbed, while making other parts available for game ranching and viewing.

Bibliography

BERE, R: *The Way to the Mountains of the Moon,* Arthur Barker, London, 1966

BERE, R: *Antelopes,* Arthur Barker, London, 1970 and Arco Publishing Company, New York, 1970

BROWN, L: *Africa,* Hamish Hamilton, London, 1965

BROWN, L: *African Birds of Prey,* Collins, London, 1970

CALDER, N: *Restless Earth,* British Broadcasting Corporation, London, 1972

CLOUDSLEY-THOMSON, J L and CHADWICK, M J: *Life in Deserts,* G T Foulis, London, 1964

COATON, W G H: "Association of Termites and Fungi", *African Wild Life,* Vol. 15, No. 1 (March 1961), p. 39

COE, M J: "The Biology of *Tilapia grahami* Boulanger in Lake Magadi, Kenya" *Acta Tropica Seperatum,* Vol. 23, No. 2 (1966)

COLE, S: "Ancient Mammals of Africa", *New Scientist,* Vol. 394 (June 1964), p. 606

COWIE, M: *The African Lion,* Arthur Barker, London, 1966 and Golden Press, New York, 1966

DORST, J: *Before Nature Dies,* Collins, London, 1970

DORST, J and DANDELOT, P: *A Field Guide to the Larger Mammals of Africa,* Collins, London, 1970

GROVES, P: *Gorillas,* Arthur Barker, London, 1970 and Arco Publishing Company, New York, 1970

HALL, R: *Discovery of Africa,* Hamlyn, London, 1970

JARVIS, J U M and SALE, J B: "Burrowing and burrow patterns of East African mole-rats *Tachyoryctes, Heliophobius* and *Heterocephalus*", *Journal of Zoology,* Vol. 163, Part 4 (April 1971), p. 451

KYLE, R: "Will the Antelope recapture Africa?", *New Scientist,* Vol. 53 (March 1972), p. 640

NAPIER, J R: "Profile of Early Man at Olduvai", *New Scientist,* Vol. 386 (April 1964), p. 86

NEAL, E: *Uganda Quest,* Collins, London, 1971

WAGER, V A: *The Frogs of South Africa,* Purnell, South Africa, 1965

WILLIAMS, J G: *A Field Guide to the National Parks of East Africa,* Collins, London, 1967

WILLIAMS, J G: *A Field Guide to the Birds of East and Central Africa,* Collins, London, 1963

WILLIAMS, J G: *A Field Guide to the Butterflies of Africa,* Collins, London, 1970

Glossary

Aestivation Dormancy during heat and drought.
Agaric Umbrella-shaped fruiting body of a fungus.
Antelope A ruminant characterized by cylindrical annulated horns; a name loosely applied to all gazelles and related forms.
Biomass Total weight of living animals to a given area.
Boundary effect The concentrating of animals where two vegetational zones meet.
Brood parasite Animal (particularly bird) that lays its eggs in the nest of another species which then rears the adopted offspring.
Browser An animal, especially an ungulate, which feeds on the twigs and leaves of shrubs and trees.
Bush Loose term for any East African countryside that is not farmland or forest; dry country with fairly dense bushes and small trees, but usually with some grass grading to wooded and open savannah in higher rainfall areas, or to semi-desert (sub-desert steppes) in lower rainfall areas; also known as Nyika ("wilderness").
Characins Very diversified group of fishes recognized by small adipose fin on the back between dorsal fin and tail; occur in Africa and tropical America.
Chelonian Group name for tortoises, terrapins or turtles.
Conidia Small fruiting bodies of fungi produced by constriction of specialized structures, conidiophores.
Convergent evolution The development of similar characters in animals and plants belonging to different groups.
Ecology The study of the mutual relationship between animals and their environment.
Feral Animal or plant that has escaped from domestication and reverted to living in the wild.
Gazelle A small to medium-sized, slender, graceful antelope, with long legs giving great speed (e.g. gerenuk, Grant's gazelle, impala).
Gymnosperm Plant whose seeds are not enclosed in a true ovary (e.g. conifers).
Hygroscopic Readily absorbing or retaining, or becoming coated with, moisture.
Individual distance The distance between individual birds of flocking species, maintained by the distance at which they can only just peck one another.
Insulberg Huge isolated granite boulder; "island rock".
Larva Immature stage of insects in which the young do not at all resemble the adults cf. **nymph** (see below).
Mimicry Resemblance in colour and/or structure as a means of camouflage or self-protection.
Nymph Immature stage of insects in which the young resemble the adults except for the absence of wings cf. **larva** (see above).
Ovipositor A specialized structure, usually a long tubular organ particularly of female insects, for laying eggs in a suitable place.
Palearctic A division of the Earth which comprises Europe, Asia north of the Himalayas, northern Arabia and Africa north of the Sahara.
Parthenogenesis Reproduction without fertilization by the male.
Plankton Floating or weakly swimming, usually very small, plants or animals in a body of water.
Protozoa Single-celled or non-cellular animal organism.
Pupa Inactive third or chrysalis stage of insects that have larvae; often enclosed in a cocoon.
Rift valley Elongate depression on Earth's surface produced by vertical displacement.
Ruminant Member of a sub-order of even-toed ungulates which chew the cud.
Savannah Tropical African grassland, open or dotted with small trees (tree savannah); corresponds to "steppe" in Asia and "prairie" in North America. In South Africa, temperate grassland is known as "veld".
Symbiosis Living together of organisms (two animals, two plants or plant and animal) in mutually beneficial partnership.
Thermal An ascending column of warm air.
Ungulate Hoofed mammal; now restricted to the Perissodactyla or odd-toed ungulates (e.g. zebras, rhinoceroses) and Artiodactyla or even-toed ungulates (e.g. hippopotamus, pigs, giraffes, antelopes).
Viviparous Giving birth to live young in contrast to laying eggs.

Index

Numbers in italics refer to illustrations